ADVANCES IN MEASUREMENT IN EDUCATIONAL RESEARCH AND ASSESSMENT

Titles of related interest

KEEVES
Educational Research, Methodology, and Measurement: An International Handbook
(2nd edn)

KEEVES & LAKOMSKI
Issues in Educational Research

ROBINSON
Problem-Based Methodology: Research for the Improvement of Practice

SCHEERENS & BOSKER
The Foundations of Educational Effectiveness

Related journals – sample copies available on request

Learning and Instruction
International Journal of Educational Research
Studies in Educational Evaluation
Evaluation and Program Planning

Advances in Measurement in Educational Research and Assessment

Edited by

Geofferey N. Masters

and

John P. Keeves

1999

PERGAMON

An imprint of Elsevier Science

Amsterdam · Lausanne · New York · Oxford · Shannon · Singapore · Tokyo

ELSEVIER SCIENCE Ltd
The Boulevard, Langford Lane, Kidlington, Oxford OX5 1GB, UK

First edition 1999

Library of Congress Cataloging-in-Publication Data
Advances in measurement in educational research and assessment/
    edited by Geofferey N. Masters and John P. Keeves. — 1st ed.
        p.   cm.
    Includes bibliographical references.
    ISBN 0–08–043348–0
    1. Educational tests and measurements. 2. Education-Research-
Evaluation.   I. Masters, Geofferey N.   II. Keeves, John P.
LB3051.A539   1999
371.26—dc21
                                                                            98–44199
                                                                            CIP

British Library Cataloguing in Publication Data
A catalogue record from the British Library has been applied for.

ISBN: 0-08-043348-0

⊗ The paper used in this publication meets the requirements of ANSI/NISO Z39.48-1992 (Permanence of Paper).

Printed in The Netherlands.

Contents

Part II: Applications of Measurement in Research and Assessment

The Contributors

Contributors are listed in alphabetical order together with their affiliations. Titles of articles which they have authored follow in alphabetical order, along with the respective page numbers. Where articles are coauthored, this has been indicated by an asterisk preceding the article title.

ADAMS, R. J. (Australian Council for Educational Research, Melbourne, Australia)
* Charting of Student Progress, 254–267.

ALAGUMALAI, S. (School of Education, The Flinders University of South Australia, Adelaide, Australia)
* New Approaches to Measurement, 23–42.

ALLERUP, P. (Danish Institute for Educational Research, Copenhagen, Denmark)
Rasch Measurement Theory, 64–84.

ANDERSON, E. B. (University of Copenhagen, Copenhagen, Denmark)
Sufficient Statistics in Educational Measurement, 122–128.

ANDRICH, D. (Murdoch University, Murdoch, Western Australia)
Essays: Equating of Marks; 176–185, Rating Scale Analysis, 110–121.

BARNARD, J. J. (Australian Council for Educational Research, Melbourne, Australia)
Item Analysis Test Construction, 195–206.

BLEISTEIN, C. A. (Educational Testing Service, Princeton, New Jersey, United States)
* Item Bias, 220–234.

KEEVES, J. P. (School of Education, The Flinders University of South Australia, Adelaide, Australia)
* Introduction, 1–19, * New Approaches to Measurement 23–42,
* Issues in Educational Measurement, 268–281.

KOLEN, M. J. (American College Testing, Iowa City, Iowa, United States)
    Equating of Tests, 164–175.

LINACRE, J. M. (University of Chicago, Chicago, Illinois, United States)
    Individualized Testing in the Classroom, 186–194.
    Judgments, Measurement of, 244–253.

MASTERS, G. N. (Australian Council for Educational Research, Melbourne, Australia)
    * Charting of Student Progress, 254–267; * Introduction, 1–19; * Issues in
    Educational Measurement 268–281. Partial Credit Model, 98–119.

ROGERS, H. J. (Teachers College, Columbia University, New York, United States)
    Guessing in Multiple Choice Tests, 235–243.

SCHEUNEMAN, J. L. (Educational Testing Service, Princeton, New Jersey, United
    States)
    * Item Bias, 220–234.

SCHLEISMAN, J. L. (University of Minnesota, Minneapolis, Minnesota, United States)
    * Adaptive Teaching, 129–137.

STOCKING, M. L. (Educational Testing Service, Princeton, New Jersey, United States)
    Item Response Theory, 55–63.

SWAMINATHAN, H. (University of Massachusetts, Amherst, Massachusetts, United
    States)
    Latent Trait Measurement Models, 43–54.

UMAR, J. (Ministry of Education and Culture, Jakarta, Indonesia)
    Item Banking, 207–219.

VAN DER LINDEN, W. J. (University of Twente, Enschede, The Netherlands)
    Computerized Educational Testing, 138–150.

WEISS, D. J. (University of Minnesota, Minneapolis, Minnesota, United States)
    * Adaptive Testing, 129–137.

WILSON, M. (University of California, Berkeley, California, United States)
    * Charting of Student Progress, 254–267.
    Measurement of Developmental Levels, 151–163.

WRIGHT, B. D. (University of Chicago, Chicago, Illinois, United States)
    Rasch Measurement Models 85–97.

1   Introduction

J. P. Keeves and G. N. Masters

Since the early 1960s there have been substantial developments in the fields of measurement in education, psychology and the social sciences. The application of these developments to problems in assessment, evaluation and research was, until the mid 1980s, limited to those research workers and graduate students who had ready access to powerful mainframe computers, which had the appropriate programs installed, or to those who had the programming skills to modify or write programs to serve their own particular purposes and for their own idiosyncratic computer system. These constraints limited the use of these developments in measurement to institutions engaged in conducting major assessment programs and to those centers in universities that were engaged in measurement research. However, these limitations did not prevent the advancement of theoretical and statistical research in the field, particularly in Europe (for example, Fischer and Molenaar, 1995; van der Linden and Hambleton, 1997). Consequently there is, in the late 1990s, a strong body of well-established theory that awaits widespread application in programs of assessment, evaluation and research where more accurate measurement is required to provide new understandings of educational achievement and educational, psychological and societal processes. Unfortunately, information on the wide variety of powerful procedures which now exist to advance measurement in these areas would appear to be hidden away in journals and reference works. Such volumes report statistical research, and while increasingly accessible, do not direct attention to the power of the advances which have been made, when taken collectively have the potential to transform many fields of inquiry. Fortunately, however, there is the emergence, in the mid 1990s, of a new range of computer programs, particularly from Australia, QUEST (Adams and Khoo, 1993), CONQUEST (Adams, Wilson and Wu, 1997), RUMM (Andrich et al., 1996) and RUMMFOLDS (Luo, Andrich and Sheridan, 1997) that are versatile, user-friendly and readily available to tackle a wide range of measurement problems that were formerly intractable.

While there is a large number of very different situations in educational, psychological and social science research to which new approaches to measurement can be applied, it is of value to list some of the more important problem situations, where advances are being made.

1 Education and psychology are concerned with stability and change in human characteristics that are influenced by learning. The lack of instruments that are independent of the items or tasks sampled in the measurement process, and that are calibrated in a way that is independent of the sample of persons used in calibration has posed problems that could only be partly overcome through the use of complex norming and equating procedures. Advances in measurement have largely eliminated these problems.

2 The measurement of attitudes and values has been seriously restricted by an inability to measure with accuracy the degree of emotional intensity with which a person responds to a particular statement or situation. The procedures of rating scale analysis overcome these limitations (Andrich, 1978).

3 Likewise in the measurement of student learning, it has not previously been possible to give credit for a partially correct response that would indicate accurately the relative difficulty of the different aspects of solving a problem or completing a task. The procedures of partial credit scaling serve to resolve these shortcomings (Masters, 1982).

4 In educational assessment a widely used procedure involves providing a student with a range of essay type questions from which the student is free to select a specified number of topics on which to write. The different topics are not of equal levels of difficulty and the scoring procedures not only require a range of score values, but must also take into consideration the relative difficulties of responding to the different topics. Essay scaling procedures have been developed which provide for the use of alternative or optional questions as well as a wide range of score values.

5 Many theories of learning and development are based on the idea of stages or discontinuities, and the testing of such theories has been limited by an inability to estimate the magnitudes of the hypothesised saltus or leaps that are believed to occur. The saltus measurement model is designed to estimate the magnitudes of discontinuities in such problem situations (Wilson, 1984).

6 In many assessment situations the persons involved do not always respond in ways that are regular or consistent. By locating the performance of a person on the same scale as the difficulties of tasks or items, it becomes possible to examine the consistency or lack of consistency in a person's responses to a series of tasks.

7 Likewise, in many assessment situations several raters are employed, and just as it is possible to place respondents on the same scale as that employed for the calibration of items or tasks, so too is it possible to place raters on that scale. In this way it is possible to calibrate the performance of raters and to make allowance for any systematic bias that might have occurred.

8 The use of procedures of measurement for the assessment of a particular

characteristic demands that characteristic is unidimensional or involves the presence of one or more traits that are operating in unison with each other. However, situations arise in practice where the requirement of unidimensionality is not met since more than one dimension is involved in the responses obtained. Scaling and scoring procedures have been advanced that provide for multidimensionality through the estimation of a profile of scores (Wilson and Adams, 1993).

These advances have occurred in both theory and through the preparation of computer programs by means of which the necessary estimation of parameters of performance can be undertaken. Widespread experience is required in the use of these estimation procedures and in the employment of the theoretical principles before the full validity or strength of the procedures can be established. It is inevitable that further theoretical and practical developments will occur.

However, in the late 1990s the challenge is to extend the widespread use of the procedures that have been developed in order to consolidate the progress that has been made and to confirm their power and validity.

The challenge to educational and psychological researchers as well as practitioners in assessment and evaluation is not only to keep abreast of the developments that have occurred, but also to learn to use them in ways that would facilitate both the conduct of inquiry as well as the improvement of practice. Nevertheless, there has been at the same time as these developments have occurred, a turning away by many in education from the use of measurement both in research and practice. This has led to widespread debate about a quantitative and qualitative divide in research methodology, as well as a polarisation in approaches to the assessment of learning outcomes between those who accept the principles of measurement and those who reject them.

1   Concerning the nature of measurement

1.1   Removing the quantitative and qualitative divide

It is frequently claimed that there are two different modes of inquiry in education, and the behavioral and social sciences which lead to distinctly different quantitative and qualitative approaches to research. Moreover, some would contend that these two approaches are two different paradigms. However, this simple dichotomy involves a serious failure to understand the nature of both quantities and qualities. Kaplan (1964) has argued that it is necessary to emphasise from the outset that measurement is not an end in itself, it merely performs an instrumental function in inquiry. He also argued that there is a danger of assuming that measures have an inherent value, without regard to the nature of the object being measured, and the intrusion of the observer into the measurement process. Furthermore, there is a tendency to disregard how the number assigned in measurement should be used in analysis. The treatment of measurement as if it had intrinsic scientific value is referred to by Kaplan (1964) as the *Mystique of Quantity*. There is, however, a more pervasive *Mystique of Quality*, which considers that any attempt which is made to measure in educational, psychological or societal research is a gross distortion and obfuscation of both objects and events. Those who adhere to this latter perspective regard qualitative methods as the only meaningful way to investigate a particular problem situation.

If these two views are considered as alternatives or even as opposites that are complementary then a serious misunderstanding of the nature of both quantities and qualities has occurred. Kaplan (1964, p. 207) has clarified the point at issue in the following statement:

> Quantities are of qualities, and a measured quality has just the magnitude expressed in its measure.

Every measurement demands some degree of abstraction. The assigning of a number to an observable characteristic or relationship requires the refinement of that characteristic or relationship before measurements can be made. This is not an assumption of measurement; it is *a requirement* that the characteristic or relationship should be accurately specified and should be unidimensional before measurement is attempted. If multidimensionality exists in the characteristic or relationship to be measured, then allowance for this known multidimensionality must be made in the scaling and subsequent analysis of the measures.

1.2 Single or multiple observations

A general distinction must be drawn between measurements that are made through a single observation or judgment and measurements that are made by combining in an appropriate way multiple observations or judgments. Errors of measurement arise from three distinct sources: (a) variability in the making of the observation or judgment due to the observer, (b) variability in the making of the observation or judgment due to the instrument being employed, and (c) variability in the characteristic being measured. Errors arising from the third source necessarily demand the making of multiple measurements by the sampling of behaviors or observable phenomena associated with the characteristic under survey. It is, however, not uncommon to make multiple observations in order to estimate and allow for observer and instrumental errors. Of necessity, such procedures lead to the combining of the multiple observations in an appropriate way.

Consequently reliance on a single observation or judgment is relatively rare in measurement in educational, psychological and societal research and practice, and the use of multiple observations is widespread. Substantial problems arise because the multiple observations must be combined. Cronbach (1960), using an analogy with the recording of music, draws attention to the distinction between bandwidth and fidelity. Thus in measurement with multiple observations it is necessary that the range of observations employed is sufficiently wide to provide a meaningful indicator of the variability in the characteristic or relationship under investigation. The range of observations is associated with the bandwidth of the recording. However, it is also necessary with multiple observations to ensure that the range of observations is sufficiently narrow to provide a high degree of fidelity and thus to satisfy the requirement that only one dimension is involved in the measurement. Only in this way is it possible to ensure that it is meaningful to combine observations. This balance between bandwidth and fidelity becomes increasingly important as advances occur in educational measurement where multiple observations are made. It is also increasingly necessary to test for unidimensionality, and where more than one dimension can be detected in the data, to employ multidimensional scaling procedures.

The same issues must be considered when multiple observations are made because of random variability that arises from the instrument being employed. Thus most observations involve error. However, the word "error" implies not a mistake, but like the knights errant of old, a wandering about a central position. Where random error is associated with measurement, then statistical procedures can be employed which, according to established conventions, enable greater precision of measurement to be obtained. It is, however, the issues of *bandwidth* and *fidelity* that must be tackled in the improvement of measurement in research and practice.

1.3   Bandwidth and fidelity

The terms, "bandwidth" and "fidelity", have some overlap in meaning with the more technical terms of "validity" and "reliability", but they are not synonymous with these more familiar terms. Moreover, they are increasingly being used in situations where validity and reliability do not suffice, since they are more directly related to the combining of observations and the making of more precise measurements.

Fidelity demands that not only should a characteristic or relationship be accurately defined, but the measurements should satisfy the requirement of unidimensionality. The development of procedures like confirmatory factor analysis provide a rigorous test for unidimensionality, that is based on the variability in the items or tasks and in the sample of respondents employed. Bejar (1983, p. 31) has, however, drawn attention to the fact that unidimensionality as tested under these conditions does not imply that only a single characteristic is involved in responding or that a single process operates. If several characteristics were to operate in unison, then unidimensionality would also hold. However, if the characteristics or processes did not operate together, then it would not be meaningful to assign numbers to any combination of the items employed, unless another operation, such as that of prediction, were to provide the rule for combination. If a set of measurements lacked this necessary fidelity, and if several identifiable dimensions were involved, then multidimensional scaling procedures would be required.

The power of measurement is that once a characteristic or relationship has been specified in detail and in a meaningful way with sufficient fidelity, then the quantities that are obtained as measures of two or more objects or events can be compared. Where measurement is made through a single observation, the issues of bandwidth and fidelity cannot apply, and a heavy load must be placed on the single observation.

The idea of bandwidth is employed to ensure that there is a sufficient range of manifestations of the characteristic being measured for the meaningful representation of that characteristic. Furthermore, under some circumstances observations can be said to supply redundant information and serve no useful purpose in measurement and must consequently be rejected. Observations can also supply information that can be shown not to be related to the specified characteristic and these observations must also be discarded. Thus decisions need to be made prior to the undertaking of measurement on the bandwidth of acceptable observations, when variability exists in the characteristic being measured.

1.4   Precision

In educational and psychological research it is common to make measurements in the form of graded responses or ratings. It is also common in the use of rating scales to assign numbers to response categories that assume equal spacing, and to assume that the error involved in rating is the same across response categories. It is also generally assumed that greater precision is obtained through the use of a greater number of response categories that have each been carefully specified. In more advanced treatments of measurement in education (Andrich, 1995) it is considered now that response categories do not have equal spacing, and that errors of assignment to response categories can not be ignored, provided the response categories employed are meaningful and have been carefully specified. In general, the larger the number of response categories the more precise the measurement. However, there may sometimes be a limit to the number of response categories that can be employed effectively in measurement by the observers making the measurement or by respondents to a rating scale.

1.5   Using numbers in measurement

The great advantage of using numbers to specify the degree or extent of a characteristic or relationship is that mathematics has provided the rules and procedures for working with numbers and matrices in order to examine relationships. In addition, statistics has provided the rules and procedures for examining the probabilities with which results might be observed or the levels of magnitude and importance of such results. It is here through the development of mathematical and statistical procedures that great advances have also occurred. Computers have made extensive computation possible, so that complex calculations which a few decades ago could not be contemplated, can now be done almost instantaneously. These mathematical and statistical procedures are necessary, not only because the interrelations between variables in educational research are complex, but also because measurements made in educational research often involve considerable error and require large samples of observations or persons for a hypothesised relationship to be detected. The rules and procedures provided by mathematics and statistics permit both the hypothesising of relationships between measured variables and the subsequent testing of these hypothesised relationships, as well as the estimation of the magnitude of specific effects.

Since the mid-1990s the computers available in the office, the classroom and the home have become so ubiquitous that they have the capacity to transform much research in education and psychology as well as the processes of instruction and learning. However, the full benefit of the calculation and estimation that can be carried out by these powerful computers is not achievable without advances in the theory of measurement. Consequently, it is of interest to examine the major developments that have occurred in recent decades in educational measurement. It is also of value to emphasise the simplicity of the ideas involved, and the ease with which computers can be employed to carry out the necessary computation. In addition, it is important to indicate the potential that these developments in both measurement and computer analysis of data have for further advances in both research and practice.

Nevertheless, there is also sometimes a marked distrust of the use of computers and the application of measurement theory. The alternative of merely employing a verbal description has greater inherent dangers, since the same words mean different things to different people, particularly in translation into different languages, now that educational and psychological research have become widespread international activities. Consequently, it is important to remember that while measurement and the computer analysis of data are not ends in themselves, there are many issues that should now be addressed using the measurement procedures and computer processing techniques that have been developed during recent decades.

Perhaps the most significant problem in education and psychology involves learning itself. That learning occurs in schools and homes is beyond doubt. However, so little is known about the factors that have actually been found to influence learning in these situations, beyond the power of time as a variable, that a major issue of concern in research and practice involves the study and advancement of learning. Progress in this field demands the more accurate measurement of learning over longer periods of time.

1.6 Validity

Much has been written about validity in educational and psychological measurement and the major review by Messick (1989, pp. 13–104) makes clear the full complexity of the issues. There are, however, two aspects of validity that are of central importance. First, there is the *meaningfulness* of the measures obtained from the measurement process. Secondly, there is the *usefulness* of the measures either in research or practice.

Kaplan (1997, p. 116) has addressed the question of validity in measurement with great clarity.

> One measuring operation or instrument is more sensitive than another if it can deal with smaller differences in the magnitudes. One is more reliable than another if repetitions of

the measures it yields are closer to one another. Accuracy combines both sensitivity and reliability. An accurate measure is without significance if it does not allow for any inferences about the magnitudes save that they result from just such and such operations. The usefulness of the measure for other inferences, especially those presupposed or hypothesised in the given inquiry, is its validity.

Thus the strength of a measuring operation and the measures it yields lies in the *simplicity* of the measures, the *correspondence* they provide between theory and the real world, their *generality* in many situations, and their *utility* both in theory and practice. These characteristics of the strength or validity of measurement which involve simplicity, correspondence, generality and utility are sustained and consolidated by the conduct of further inquiry, without which the measuring operation dies.

The operation of measurement does not necessarily imply *objectivity*, and Messick (1989, p. 58) argues that it is essential to consider the consequential validity concerned with the value implications of the measures, their associated constructs, the interpretation given to test scores, and the use made of the measures in inquiry. Kaplan has also addressed succinctly the issues associated with values and bias and their roles in measurement and inquiry.

> Inquiry itself is purposive behavior and so is subject to behavioral interpretation. The interpretation consists in part of specifying the values implicated in specific processes of conceptualization, observation, measurement, and theory construction. That values play a part in these processes does not in itself make the outcomes of these processes perjoratively subjective, nor otherwise invalidate them. A value which interferes with inquiry is a bias. Not all values are biases; on the contrary, inquiry is impossible without values. (Kaplan, 1997, p. 188.)

2   Advances in the theories of measurement

Nearly 100 years ago in 1905, Binet and Simon published the report of their initial scale for the assessment of intelligence. At approximately the same time Spearman published his model for test scores that laid the foundations of classical test theory. This theory, with its concepts of true score, measurement error and the index of test reliability dominated the field of educational and psychological measurement for approximately 50 years. In time a substantial body of statistical theory and related computational techniques was assembled. Nevertheless, two major issues emerged over time as significant problems. The first involved the estimation of the contribution of different sources of error to the total error variance associated with the use of a test. The second was related to the selection of items or tasks that were included in a particular version of a test, since most tests employed in education were formed as a combination of items or tasks. Although parallel forms could be developed for tests, in general, the selection of items was left to the judgment of test constructors or research workers who developed a particular instrument.

Gradually, two other competing theoretical approaches to educational and psycho-logical measurement have evolved. The theories involved have become known as *generalizability theory* which uses what might be referred to as a random sampling model for the selection of items or tasks, and *item response theory* for which a range of statistical models has been developed, but not necessarily based on the use of the logistic transformation. Generalizability theory remains relatively close to classical test theory, and is clearly oriented towards the obtaining of a total score. However, it differs from classical test theory in so far as the items employed are, in general, selected randomly from a pool of items. This enables the different components of variance to be calculated as associated with variability between items, raters and persons, and enables reliabilities to be estimated for these different facets of error.

Item response theory is primarily concerned with the probabilities associated with the level of performance of an individual relative to a particular item. This leads to major differences between the two theories in the specification of the content being tested. In generalizability theory there is a pool of homogeneous items from which a sample of items can be drawn. Alternatively, it is possible to stratify items by type and content into a hierarchy of levels of difficulty from which sampling can occur. In item response theory, it is necessary to ensure that the items are located along a single latent trait continuum, that is invariant across the groups of students to whom the test is given. However, provided the items satisfy this condition of unidimensionality the specific items employed in a test need not be randomly selected.

A further important difference between generalizability theory and item response theory, is that in generalizability theory the estimates of errors of measurement and the reliability of a test apply to the test as a whole. However, in item response theory it is possible to assign an estimate of error to each person and each item at each level of the scale continuum.

In classical test theory a correction for guessing on multiple choice items may be applied. However, this correction is likely either to under-estimate or over-estimate the extent to which guessing might have occurred on particular items, although it does make some allowance for the extent to which individuals might differ in their tendency to guess. Furthermore, in the three-parameter logistic model in item response theory an allowance is made for guessing with respect to each individual item, with the assumption that guessing is a characteristic of items and not persons.

Criterion-referenced measurement involves another theory of test construction and use which permits the interpretation of the performance of a person taking a test to be related to well-defined objectives (Popham, 1978). There is, in addition, a type of criterion-referenced test referred to as a *mastery test* which is based on models of school learning (Carroll, 1963; Block, 1971; Bloom, 1974). However, both of these approaches to test development encounter problems in the selection of items and the determination of an appropriate standard across all items in a test for the assignment of mastery, or for achieving a criterion level of performance.

Classical test theory has been employed, traditionally, in the construction of achievement tests. It may be referred to as "weak true score" theory since it merely involves two assumptions that: (a) the observed raw scores can be decomposed into two additive components, namely, true score and error; and (b) repeated estimates of true scores for a person are not linearly correlated with the error scores for that person. Keats (1997) has argued that "strong true score theory" which employs an appropriate statistical model awaits development, and when such development has occurred it could well replace existing procedures. However, towards the end of the twentieth century it is "weak true score theory" that remains the most widely used approach for the construction and scoring of tests, as well as for other types of instrumentation in educational research and practice. Nevertheless, weak true score theory presents many problems for the user including the dependence of the reliability estimate on the variability in the sample that is tested, as well as the deviation of the scale of scores from an interval scale, although with a large number of items the scale would seem to approximate satisfactorily to an interval scale. The main problem associated with the use of generalizability theory is that rarely is there a large pool of appropriate test items from which sampling could occur in the construction of a test.

3   Conjoint measurement and item response theory

One of the major problems in measurement in education and in the social and behavioral sciences is that there is an interaction between the person being measured and the instrument involved in measurement at the time measurements are made. As a consequence the performance of a person is not independent of the measuring instrument employed. This uncertainty or confounding that arises between the person and the instrument is circumvented by the procedure of conjoint measurement advanced by Rasch in 1960. In conjoint measurement it is always the performance of a person relative to a particular item that is being considered in terms of probabilities. Thus a person's ability is set at the same level as the difficulty of an item if that person has a specified probability, commonly 50 per cent, of responding correctly to the item.

Furthermore, clear benefits are gained in educational measurement through the location on a common scale of the students measured by an instrument and the items in that instrument. In this way the idea of a criterion level of performance is replaced by the location of a student's level of performance with respect to particular items. Specific levels or standards of performance on a scale of this type can also be stated and shown in terms of either the characteristics of the items or the characteristics of the students or alternatively by a defined level on the scale, once a fixed point has been set to anchor the scale with respect to items or to students.

Rasch (1960) also proposed that the logistic function should be employed to construct an interval scale that related the position of an item on a scale, which was associated with the characteristic being measured, to a student's probability of success on that item. In the scale of conjoint measurement, when the level of performance of a student just matched the difficulty level of an item then the student would have a one to one chance or a 50 per cent probability of answering the item correctly or incorrectly.

The Rasch model is then constructed as a logistic function of the odds associated with the performance level of a person with respect to the difficulty level of an item. The logistic transformation of the response odds also enables the item and person parameters and their estimated values to be expressed as separate components on the conjoint scale. Conditional probabilities are used to separate the person parameters from the item calibration, and the item parameters from the estimation of the person parameters. The necessary requirement for this estimation procedure to provide meaningful results is that both the items and the persons must fit a unidimensional model and behave in a consistent way across different samples. This requirement simply imposes a restriction, similar but not completely identical to that used in classical test theory, that should be satisfied before it is considered appropriate to add together item responses to obtain a total score.

In achievement testing using classical test theory, such a restriction has been largely applied through item analysis and item selection, although a sufficiently large number of items has commonly been employed to prevent non-conforming items from having a damaging effect on a total score. In attitude and descriptive scale construction it has been an accepted practice for items to be eliminated from a scale if they did not satisfy this requirement for the additivity of scores. The more restrictive requirement of Rasch scaling is, however, a small price to pay for the advantages of measurement that are gained by using the logistic function to construct a scale of performance.

It should be noted that the Rasch scale is not only an interval scale, but also has its own natural metric, with the scale unit referred to as a logit. All that is required is that a fixed point should be specified in order to determine the location of the scale. In addition, the errors involved in the estimation of both the item difficulty parameters and the person performance parameters are obtained for each individual item and person, rather than for the instrument as a whole, provided conditions of independence of items and persons are maintained. This scaling procedure eliminates the dependence of the item parameters on the sample of persons used in calibration, and the dependence of the person parameters on the sample of items used. Under these circumstances, the scale so formed has some properties of an absolute scale, but fails to have an absolute zero, being an interval, but not a ratio scale. Such a scale has substantial advantages in educational and psychological measurement, with the only requirement that the items and the persons used in the calibration of the scale must satisfy the condition of unidimensionality (see *Rasch Measurement Theory*).

3.1 *Advantages of Rasch measurement*

Some of the immediately obvious advantages of the Rasch scale in educational and psychological measurement can be listed. First, in the equating of different instruments that are known to measure the same student trait or the same underlying dimension of performance, all that is required is the appropriate location of the zero points of the two scales in terms of the relative average difficulty levels of the items. The ease with which the equating of instruments (see *Equating of Tests*) on the same scale can be carried out, and the ease with which it is possible to test whether two instruments measure the same dimension, is of great advantage in the investigation of learning.

Second, when two subgroups drawn from a common population are administered the same calibrated instrument, they may differ considerably in their mean level of performance. However, the relative positions of the difficulty parameters of particular items on the calibrated scale would only change if items were biased with respect to the two groups (see *Item Bias*). The detection of this type of biased items provides information of value in education, because the existence of bias reflects either differences in the learning experiences involved for providing a correct response to the item, or deficiencies in the construction of an item so that it would favor one particular group to the disadvantage of the other group. The term, "differential item functioning", is now commonly employed for items that behave differently for different subgroups.

Third, the location of persons on the same scale as items permits the ready identification of the inconsistent behavior of persons. A particular person might be expected to respond correctly to those items where the difficulty level is well below that person's performance level, and fail to respond correctly to those items where the difficulty level is well above that person's performance level. This also permits the accurate diagnosis of problems in student learning since items can be readily identified that a particular person is expected to get correct but fails to answer correctly.

Fourth, while the tendency for a student to guess responses could have damaging consequences for the unidimensionality of a scale, if students were advised not to guess at random in responding to multiple choice test items, then their scores could be accurately calculated using the information available on the set of items to which they responded. However, there would be the need for a student not to have omitted a subset of items that reflected a particular content bias. If these restrictions were satisfied, then the problem of guessing would be largely avoided.

Finally, in the estimation of a person's performance on a specific characteristic, it would be possible to select from a pool of calibrated items a subset of items which were close to that person's level of performance and to use a sufficient number of items in that subset to estimate that person's performance to a specified degree of accuracy. A computerised testing procedure, referred to as *computer adaptive testing*, (see *Adaptive Testing*) has been developed to facilitate the estimation of a person's performance without subjecting that person to a lengthy test containing a very large number of items that adds little to the accuracy of estimation.

Some of these advantages of the Rasch model are shared with the two and three parameter item response models (see *Item Response Theory*) which introduce an item discrimination parameter and guessing parameter respectively into the model. The use of these alternative models requires large samples of persons for calibration and the estimation of these parameters of the model is no longer independent of the sample of persons or the sample of items employed. The requirement that all items and persons involved in scale calibration should conform to the Rasch model has the benefits of simplicity and generality that are lost when the two and three parameter models are used. However, a price is paid for the use of the Rasch model, that involves the exclusion from calibration of non-fitting items and persons. Estimates of person performance may nevertheless be made for those persons excluded, and advantages are gained through improved measurement.

3.2   The strengths of the Rasch model

The strengths of the Rasch model lie in the simplicity of the algebra involved as well as in the extension of the model to cover a range of situations. No longer is unidimensionality a restriction, provided a limited number of dimensions have been hypothesised, and the items and persons are constrained to these dimensions. There is no place for noise or for items and persons that do not conform to the dimensions specified in a model. If the necessary requirements of the model are satisfied, the benefits are substantial as a result of the shift from deterministic approaches to measurement to probabilistic or stochastic models in order to advance the accuracy of measurement. There remain some limitations associated with a ceiling or floor for a particular instrument when respondents answer correctly all items or no items respectively. However, the ceilings or floors are false in so far as the use of further items that conform to the unidimensional scale involved would permit the accurate estimation of performance over a wider range. Thus, unlike classical test theory, where the test or instrument is the scale, in Rasch scaling the scale is independent of the items in the test and the sample of persons employed in calibration.

It must be recognised that problems are encountered if a latent trait scale does not remain invariant over the population being investigated. However, the capability exists to develop a scale that has the property of invariance, as well as to construct multidimensional scales. Insufficient work has been carried out into the use of multidimensional scales to determine the limitations imposed on the research questions that can be meaningfully investigated. In addition, further research is required into the robustness of scales measuring educational achievement in contrast to tightly defined cognitive abilities, as well as the validity of the results obtained under different circumstances.

4   Assessment, evaluation and measurement

The ideas of assessment, evaluation and measurement create much discussion in education. It is important to recognise that these three terms need to be interpreted separately and differently. However, in the minds of many researchers and practitioners these terms together with the topic of testing, would seem to be used interchangeably. What assessment, evaluation and measurement have in common is testing. Each frequently, but not always, makes use of tests. Nevertheless, none of these terms is synonymous with testing, and the types of tests required for each of the three processes may be very different. The three processes are considered below in reverse order.

4.1 Measurement

The regular dictionary definition of "measurement" as 'assigning a numerical quantity to ...' serves well in most applications of educational and psychological measurement. While instruments such as rulers and stopwatches can be used directly to measure height and speed, many characteristics of educational interest must be measured indirectly. Thus, ability tests are typically used to measure such characteristics as intelligence, and achievement tests are used to measure the amount of knowledge learned or forgotten. The items employed in these tests are manifestations of a latent variable. They are not the characteristic itself. Latent trait test theory, or as it is now known, item response theory, recognises this separation and seeks to estimate performance on the underlying latent characteristic. Likewise, classical test theory also consideres a *true score* as distinct from a *raw score*. However, classical test theory can only estimate a true score by using the group properties of a test, the reliability and standard error of measurement. Item response theory does not need to employ such concepts as reliability and standard error of measurement, in the measurement of individual performance, since errors of estimation are calculated for each person separately. It is, nevertheless, apparent that measurement is not undertaken as an end in itself. It is a useful operation in the processes of evaluation, or for research where characteristics must be measured, or as part of the tasks of assessment of student performance.

4.2 Evaluation

In general, the use of the term "evaluation" is reserved for application to abstract entities such as programs, curricula and organisational situations. Its use implies a general weighing of the value or worth of something. Evaluation commonly involves making comparisons with a standard, or against criteria derived from stated objectives, or with other programs, curricula or organisational situations. Evaluation is primarily an activity involved in research and development. It may require the measurement of educational outcomes, and it may involve the testing of both individuals and groups. Its potential importance in the improvement of educational practice is widely recognised, but fierce controversy surrounds the issue of the methods that should be used and the part that measurement should play in the conduct of an evaluation. Indeed it is possible for an evaluation to be conducted that does not involve any measurement of observable characteristics and that makes judgments against specified criteria in a holistic way which involves a systemic approach to evaluation. Most judgments of an evaluative kind that are made in education would seem to be holistic in nature and to be based on a global examination of a situation.

4.3   *Assessment*

In general, the term "assessment" should be reserved for use with reference to people. It may include the administration of tests, or it may simply involve activities of grading or classifying according to some specified criteria. Student achievement in a particular course might be assessed, or students' attitudes towards particular aspects of their schooling might be examined. Such assessments are commonly based on an informal synthesis of a wide variety of evidence, although they might include the use of test results, or responses to attitude scales and questionnaires.

Attention is increasingly being given to improving the quality of assessment, through the systematic specification of levels of performance. However, such assessments can be converted to a scale of measurement, through the use of the partial credit model (see *Partial Credit Model*) or the scaling of essay marks (see *Essays: Equating of Marks*), and there is little need to view assessment as a process that does not involve measurement. It is, nevertheless, unfortunate that the term student evaluation is now being widely used as a consequence of the growing emphasis on the evaluation of educational programs and the financial support made available for such work. The use of the term student evaluation implies making a value judgment on the performance of a student that involves a consideration of the student's worth relative to other students. This is unnecessary and undesirable, because the development of a scale of achievement, permits student learning to be examined, the responsibility for which is shared by the student, the school and the home. No longer is it necessary to consider a student's worth relative to other students, if the emphasis in education is on the facilitation of student learning and the attainment of standards of performance. Such an approach to the assessment of student performance would change, in a radical way, traditional procedures based on selection and competition.

5   The need for a volume concerned with advances in measurement and assessment in education

The discussion in the earlier sections of this introduction points out that there are two main approaches to measurement in the fields of education, psychology and the social sciences. These two approaches are built around classical test theory and item response theory. The traditional approach to measurement, which involves classical test theory, is still very widely employed, and is likely to continue as the basis for the construction of tests and instruments for assessment, evaluation and research, because of the simplicity of calculating and using a total score. However, item response theory has so many advantages in situations where the power of an interval scale is required and where the shortcomings of classical test theory limit the nature of the comparisons involving measured quantities that can be made. At this time, and for the foreseeable future, it is not a choice between the use of one theory or the other, but rather their use according to circumstances and the nature of the problem situation under investigation. In the past the use of item response theory has been seriously limited by the problems of computation and the lack of computer programs being readily available to perform the necessary calculations. The past decade has seen major advances in the range of problem situations that can be meaningfully investigated using item response measurement, and the ease with which the necessary computer analyses can be undertaken. As a consequence, educational and psychological practitioners as well as research workers need to be familiar with both theoretical approaches to measurement.

This volume seeks to be a truly international publication by providing coverage of both approaches and by drawing on the writings of scholars from different parts of the world, as well as looking forward to the future. It is clear that measurement in the fields of education, psychology and the social sciences is a rapidly expanding one, although pockets of opposition continue to exist in Britain and North America. This volume aims at capturing the changing nature of the approach to problems of measurement, and advancing reform not through the development of new theory, but through the widespread application and use of the theory and procedures that have evolved during the past decade.

The chapters in this volume are presented in two major sections:
(a) theoretical developments, and
(b) applications of new developments.

The first section addresses more specifically the theoretical developments that have occurred, and because of the strong measurement properties of the one parameter logistic or Rasch model, the emphasis in the section is primarily on the developments that have occurred in relation to this model. These developments are now so extensive that it is probably no longer appropriate to speak of the Rasch model, but rather of the Rasch family of models, many of which were foreseen by Georg Rasch in 1960. As a consequence, there is now a substantial body of modern item response theory that is being applied increasingly across the world for the tasks of measurement in the fields of education, psychology and the social sciences. The second section considers many of these applications, including the growing use of computers in the assessment of student learning as well as the important developments associated with the measurement of change over time with respect to student achievement, and the setting and attainment of standards in school performance. The closing article in the volume is concerned with issues in educational measurement that must be addressed in future developments.

Nevertheless, the widespread use of item response theory has been restricted by the failure of many researchers and practitioners in the fields of education and psychology, to understand not only the ideas but also the power of the application of these ideas to both familiar and new problems. Thus this volume not only addresses the theoretical ideas, but also the application of these ideas in order to help advance their use more widely.

6   The preparation of this volume

This volume has been developed from the first and second editions of the *International Encyclopedia of Education* which were edited by Torsten Husén and Neville Postlethwaite with the assistance of an editorial board and that were published in 1985 and 1994 respectively. Both editions of the Encyclopedia were organised around areas of scholarly specialisation related to education. Two areas involved Educational Research Methodology and Educational Measurement. For the first edition of the Encyclopedia, the section on Educational Research Methodology was edited by J. P. Keeves and the second section on Educational Measurement and Evaluation was edited by the late B. H. Choppin. For the second edition the topics on Educational Measurement were subsumed within the general field of Educational Research, Methodology and Measurement with J. P. Keeves as the editor. This present volume draws articles on measurement from both editions of the Encyclopedia where appropriate, and adds new articles to provide for the developments that have occurred during recent years. Thus, for this volume some new articles have been prepared and many authors have revised their entries from the time when they were first written. Therefore efforts have been made to ensure that the material presented in this volume is as relevant as possible to the field in the late 1990s and beyond.

7   How to use this volume

While it is the editors' contention that there is a unity and coherence running through this volume concerned with the theoretical developments that have occurred in educational and psychological measurement and their application to a wide variety of problems, this volume must be seen as a collection of articles, written by different scholars with different theoretical perspectives and drawn from many different countries with different research traditions. As a consequence, the volume is intended as a source book for university and college teachers to use in the preparation of lectures in courses on advanced educational measurement, for graduate students to use as a reference book for a first introduction to a measurement problem, and for practising research workers and measurement or assessment workers in examination agencies to obtain information on the developments that have occurred over recent years which they might employ in their work. It is clear that no individual article is complete in itself. Thus every article seeks not only to be relevant and up to date, but also to provide guidance, through a concise set of references and a bibliography to key articles or publications likely to be readily available from which the scholar, student, teacher or research worker could obtain further information. Furthermore, in order to facilitate the search for information, references are provided within each article to other articles in the volume where related information has been presented.

In addition, both a detailed *Name Index* and a *Subject Index* based on key words and phrases have been compiled to assist the reader in the search for information. As a consequence, the user of this volume who wants information on a specific topic could begin by looking up appropriate words in the *Subject Index* in order, either to locate an entry related to the topic, or to identify entries where the topic, as referenced by the key words, is considered. In a similar way, the *Name Index* can be used to locate references to the writings of a particular author who is known to have made a substantial contribution to a sphere of research related to the topic on which information is sought. In order to facilitate this task, page numbers are given both for the bibliographic reference and for the point at which the reference is cited in the text.

Acknowledgments

No work of this nature could be prepared and published without considerable effort by many people. To several of these people a special debt of gratitude is due. First, we are grateful to Neville Postlethwaite and Torsten Husén who guided the preparation of the first and second editions of the *International Encyclopedia of Education*. Second, we are grateful to the many authors who prepared entries, carefully checked galley proofs, revised their articles, and in many cases acted as consultants for other articles in the volume. Third, we are grateful to Barbara Barrett, Editorial Director of the *Encyclopedias*, who was assisted in the preparation of the second edition by Michele Wheaton as well as to Tony Seward and David Lamkin who coordinated work on the production of this volume. Finally, we are grateful to the many who assisted with typing and editing articles in connection with the preparation of this volume. To them all our sincere thanks are offered.

References

Adams, R. J. and Khoo, S. T. 1993. *QUEST: The Interactive Test Analysis System*. Australian Council for Educational Research, Melbourne.

Adams, R. J., Wilson, M. and Wu, M. 1997. CONQUEST. Australian Council for Educational Research, Melbourne.

Andrich, D. 1978. A rating formulation for ordered response categories. *Psychometrika* **43**, 561–73.

Andrich, D., Luo, G. and Sheridan, B. 1996. RUMM, *Rasch Unidimensional Measurement Models*. Murdoch University, Perth.

Bejar, I. I. 1983. *Achievement Testing: Recent Advances*. Sage, Beverly Hills, California and London.

Block, J. H. 1971. *Mastery Learning, Theory and Practice*. Holt, Rinehart and Winston, New York.

Bloom, B. S. 1974. Time and learning. *American Psychologist* **29**, 682–8.

Carroll, J. B. 1963. A model of school learning. *Teachers College Record* **64**, 723–33.

Cronbach, L. J. 1960. *Essentials of Psychological Testing*. Harper and Row, New York.

Fischer, G. J. and Molenaar, I. W. (eds.) 1995. *Rasch Models Foundations, Recent Developments, and Applications*. Springer-Verlag, New York.

Kaplan, A. 1964. *The Conduct of Inquiry*. Chandler, San Francisco, California.

Kaplan, A. 1997. Scientific Methods in Educational Research. In: J. P. Keeves (ed.) *Educational Research, Methodology and Measurement*. Pergamon, Oxford. pp. 112–119.

Keats, J. A. 1997. Classical Test Theory. In: J. P. Keeves (ed.) *Educational Research, Methodology and Measurement*. Pergamon, Oxford. pp. 713–719.

Masters, G. N. 1982. The Rasch model for partial credit scoring. *Psychometrika* **47**, 149–74.

Messick, S. 1989. Validity. In: R. L. Linn (ed.) *Educational Measurement. Third Edition*. American Council on Education. Macmillan, New York.

Luo, G., Andrich, D. and Sheridan, B. 1997. RUMMFOLDS, *A computer program*, Murdoch University, Perth.

Popham, W. J. 1978. The case for criterion referenced measurements, *Educational Researcher* **7**(11), 6–10.

Rasch, G. 1960. *Probabilistic Models for Some Intelligence and Attainment Tests.* (reprinted 1980), University of Chicago Press, Chicago.

van der Linden, W. J. and Hambleton, R. K. (eds) 1997. *Handbook of Modern Item Response Theory.* Springer-Verlag, New York.

Wilson, M. 1984. *Measuring Stages of Growth. A Psychometric Model of Hierarchical Development.* Australian Council for Educational Research, Hawthorn, Victoria.

Wilson, M. and Adams, R. J. 1993. Marginal maximum likelihood estimation for the ordered partition model. *Journal of Educational Statistics* **18**(1), 69–90.

Part I

New Theoretical Developments

2   New Approaches to Measurement

J. P. Keeves and S. Alagumalai

The past 50 years has seen not only the consolidation of classical test theory and a growing recognition of its shortcomings but also the emergence of new approaches to educational and psychological measurement. These new approaches have not gained widespread recognition, in part, because of the dominance of the field by testing agencies in the United States, and partly because of the emerging opposition by many educators and psychologists to the use of quantitative methods. However, the developments that have occurred through the use of microcomputers for the processing of data have eliminated the statistical and technical complexities involved in employing these new approaches. As a consequence the capacity to use these approaches has shifted from the domain of the testing agency to the desktops of many university lecturers, research workers, graduate students and classroom teachers.

The basic ideas employed in these new approaches are simple to understand requiring nothing more than a knowledge of high school mathematics and statistics. This article provides an overview of these ideas as an introduction to the second section of this volume which presents these new approaches in greater detail. These new approaches have come to be known as item response theory, although the older names of latent trait theory and characteristic curve theory are still sometimes used, and are sometimes considered to describe more meaningfully the essential features of this theory.

1   The fundamental principles of item response theory

The three basic ideas employed in item response theory involve:

(a) the relativity principle, referred to as *conjoint assessment*;
(b) the probability principle, that rejects a deterministic approach to responding to test items or statements; and
(c) the measurement principle, that requires the use of an interval scale which is not restricted to a limited range, but extends without limit in both positive and negative directions.

Collectively these three ideas involve *conjoint stochastic measurement*. Each of these three ideas is explained in greater detail below. However, the use of these ideas only emerged slowly. Thurstone (1927) would appear to have been one of the first measurement theorists to recognise the importance of a probabilistic or stochastic approach to measurement. Although Guttman's (1944) highly deterministic view was subsequently advanced and gained strong endorsement, it provides a rigorous framework for the formulation of measurement principles.

Ferguson (1942) and Lawley (1943) advanced the use of the conjoint principle for the performance of a person relative to the difficulty of the task and also proposed the use of the normal distribution function to model probabilities, while Lawley (1943) showed how estimation could proceed using a maximum likelihood procedure. Nevertheless, this work faltered because of the computational difficulties involved in the estimation of the parameters of the model and the mathematical complexities of working with the normal distribution function.

It was Georg Rasch (1960), a Danish mathematician, who first used the logistic function to transform the expressions advanced by Ferguson and Lawley in Scotland in order to simplify the formulation and estimation of student performance on reading achievement tests. The use of the logistic function which was proposed by Rasch was extended by Birnbaum (1968) in Lord and Novick (1968) as an approximation to the normal distribution to incorporate test items that did not satisfy the restrictive requirements associated with the use of the Rasch model. Nevertheless, the strong measurement properties of the logistic model proposed by Rasch warrant the use of this model and the rejection of items or tasks that do not conform to the model in order to develop a unidimensional scale that measures an identifiable latent trait with a pattern of responses that approximates the pattern prescribed by a Guttman scale. Moreover, the Guttman response pattern is consistent with the use of the logistic function as a model of response distribution.

2   The relativity principle

Classical test theory focuses on the assessment of the performance of a person on a prescribed set of test items. However, there is necessarily an interaction between each person and each item in educational and psychological assessment that gives rise to the response of the person to the item in a manner that implies that the basic unit of measurement is not the person or the item, but rather the performance of the person realtive to the item. Thus if β_n is an index for the underlying ability of person n on the attribute or trait being measured, and if δ_i is an index for the underlying difficulty or facility level of the item or task i which relates to the attribute or trait being measured, then it is not β_n or δ_i that is the unit of measurement but rather the difference between the ability of the person relative to the difficulty of the item or $(\beta_n - \delta_i)$ that must be taken into consideration. Alternatively the ratio of ability to difficulty could be used. If the ability of the person exceeded the difficulty of the item, then the response would be expected to be correct or favorable, and if the ability of the person were less than the difficulty level of the item, then the response would be expected to be incorrect or unfavorable. The ability of the person and the difficulty of the item must be considered to be joined or conjoint in all analyses of responses and a principle of relativity of the person with respect to the item must underlie the task of measurement. This principle overcomes the problems that were raised in earlier decades and that claimed that measurement was not possible in the social and behavioral sciences.

3   The probability principle

The response of a person to a particular item or task is never one of certainty. In educational situations such responses are influenced by both carelessness and guessing, but even more important is the inevitably probabilistic nature of the responses of people to the situations they encounter that must be taken into consideration in measurement. Error involves, as for the knights errant of olden times, a wandering about a mean position that arises from the chance fluctuations that occur in human behavior. As a consequence a stochastic or probabilistic approach must be employed. This is, of course, not inconsistent with the use of the ideas of true score and error in classical test theory. However, in Rasch measurement, probabilities are introduced through consideration of the odds that a person gives a correct or favorable response to an item.

Thus if the ability of person n is given by θ_n and the difficulty of an item i is given by Δ_i then

if $\theta_n > \Delta_i$ the person is expected to respond correctly

or if $\theta_n < \Delta_i$ the person is expected to respond incorrectly to the item.

Moreover if the odds of a person's response to an item are given by

$$\frac{\theta_n}{\Delta_i} > 1, \text{ a correct response is expected}$$

or if, $$\frac{\theta_n}{\Delta_i} < 1, \text{ an incorrect response is expected}$$

and if, $$\frac{\theta_n}{\Delta_i} = 1 \text{ then there is a 50 per cent chance of a correct response.}$$

If p_{ni} is the probability of a correct response, then $1 - p_{ni}$ is the probability of an incorrect response, and the odds for a response are given by

$$\frac{\theta_n}{\Delta_i} = \frac{p_{ni}}{1 - p_{ni}} \tag{1}$$

and if $\dfrac{\theta_n}{\Delta_i} = 1$, then $p_{ni} = 0.5$.

4 The measurement principle

Probabilities of a correct response both for persons and for groups lie in the range of 0 to 1.0, and these constraints prevent data expressed in terms of proportions correct either for persons or items lying along an interval scale. As a consequence all attempts at measurement in terms of proportions correct or raw scores cannot be considered to form an interval scale. However, if tests are constructed so that students do not attain the ceiling or the floor and the mean score is near the middle of the range, then the raw scores and proportions obtained from such tests with large numbers of items often approximate to an interval scale.

These problems can be overcome by simply using the logistic transformation. This involves taking the natural logarithm of the odds presented in Equation (1).

Thus

$$\ln \frac{\theta_n}{\Delta_i} = \ln \left(\frac{p_{ni}}{1 - p_{ni}} \right)$$

and

$$\ln \theta_n - \ln \Delta_i = \ln \left(\frac{p_{ni}}{1 - p_{ni}} \right)$$

For simplicity, let $\ln \theta_n = \beta_n$ and $\ln \Delta_i = \delta_i$, then

$$\beta_n - \delta_i = \ln \left(\frac{p_{ni}}{1 - p_{ni}} \right)$$

and

$$\frac{p_{ni}}{1 - p_{ni}} = e^{(\beta_n - \delta_i)} \text{ or } \exp(\beta_n - \delta_i)$$

from which the probability of a correct response $(x_{ni} = 1)$ is

$$p_{ni} = \frac{\exp(\beta_n - \delta_i)}{1 + \exp(\beta_n - \delta_i)} \tag{2}$$

and the probability of an incorrect response

$$p_{ni}(x_{ni} = 0) = 1 - p_{ni}(x_{ni} = 1) = \frac{1}{1 + \exp(\beta_n - \delta_i)}$$

The great advantage of the use of the expression $\exp(\beta_n - \delta_i)$ is that the terms for person ability and item difficulty are additive and the subsequent mathematics is greatly simplified.

5 The Rasch scale

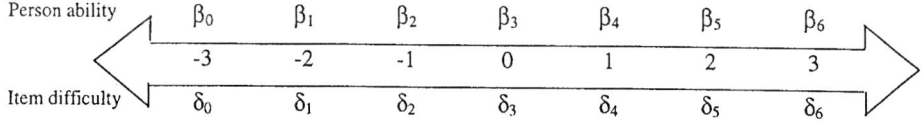

Figure 1 The Rasch scale

This formulation permits persons (β_n) and items (δ_i) to be located on the same number scale with scale values $-\infty$ to $+\infty$ assigned in such a way that an interval scale ranging from large negative values to large positive values is formed. The requirement that must be satisfied by the values assigned to persons (β_n) and items (δ_i) is that they must be related by the expression given in Equation (2)

$$p_{ni} = \frac{\exp(\beta_n - \delta_i)}{1 + \exp(\beta_n - \delta_i)}$$

where p_{ni} is the probability of a correct response given by person n to item i.

This relationship can be seen in diagrammatic or graphical form by plotting the values for

(a) items across the range of persons from scale values -3 to $+3$, and
(b) persons across the range of items from the scale values -3 to $+3$.

Table 1 presents the probabilities that are calculated by substituting the different values of β_n and δ_i in Equation (2).

Thus for Person 3 $(\beta_3 = 0)$ and Item 3 $(\delta_3 = 0)$

$$p_{33} = \frac{\exp(0 - 0)}{1 + \exp(0 - 0)} = \frac{1}{1 + 1} = 0.5$$

Table 1   Probabilities calculated for persons and items

Persons	Item 0 $\delta_0 = -3$	Item 1 $\delta_1 = -2$	Item 2 $\delta_2 = -1$	Item 3 $\delta_3 = 0$	Item 4 $\delta_4 = +1$	Item 5 $\delta_5 = +2$	Item 6 $\delta_6 = +3$
Person 0 ($\beta_0 = -3$)	0.50	0.27	0.12	0.05			
Person 1 ($\beta_1 = -2$)	0.73	0.50	0.27	0.12	0.05		
Person 2 ($\beta_2 = -1$)	0.88	0.73	0.50	0.27	0.12	0.05	
Person 3 ($\beta_3 = 0$)	0.95	0.88	0.73	0.50	0.27	0.12	0.05
Person 4 ($\beta_0 = +1$)		0.95	0.88	0.73	0.50	0.27	0.12
Person 5 ($\beta_0 = +2$)			0.95	0.88	0.73	0.50	0.27
Person 6 ($\beta_0 = +3$)				0.95	0.88	0.73	0.50

and for Person 1 ($\beta_1 = -2$) and Item 2 ($\delta_2 = -1$)

$$p_{12} = \frac{\exp(-2+1)}{1+\exp(-2+1)} = \frac{0.37}{1+0.37} = 0.27$$

If persons and items satisfy the relationship given in Equation (2) then the item characteristics curves for each item must be parallel when probability of success on the item is plotted against person ability and the mirror image person characteristic curves must also be parallel. When $p = 0.5$ the slope of the item characteristic curves must be unity ($+1.0$) and the slopes of the person characteristic curves must be -1.0. It should be noted that the scales both for persons and for items are interval scales since they are formed from the number scale shown in Figure 1. The only arbitrary property of the scale is the location of the zero point. However, once a standard group of persons or a standard set of items has been identified and a zero or fixed point chosen, the scale so formed is independent of both persons and items. The units of the scale are referred to as logits and are the natural scale units.

The only requirement that both persons and items must satisfy is that they conform to the logistic function given in Equation (2), or in visual terms the characteristic curves of both persons and items must conform to the person and item characteristic curves shown in Figure 2. Conformity to the item and person characteristic curves is a small price to pay for strong measurement and the theory of this approach is sometimes referred to as characteristic curve theory. However, the price to be paid may involve the rejection of items and persons in the calibration of the scale.

6   Extensions of the simple Rasch model

The simple Rasch model that involves a parameter for the ability or performance of persons β_n, and for the difficulty of items or tasks (δ_i) can be readily extended to include parameters for rater severities (ρ_k), thresholds in rating and partial credit scales (τ_l), and stages in development (γ_m) by the extension of the conjoint expression to

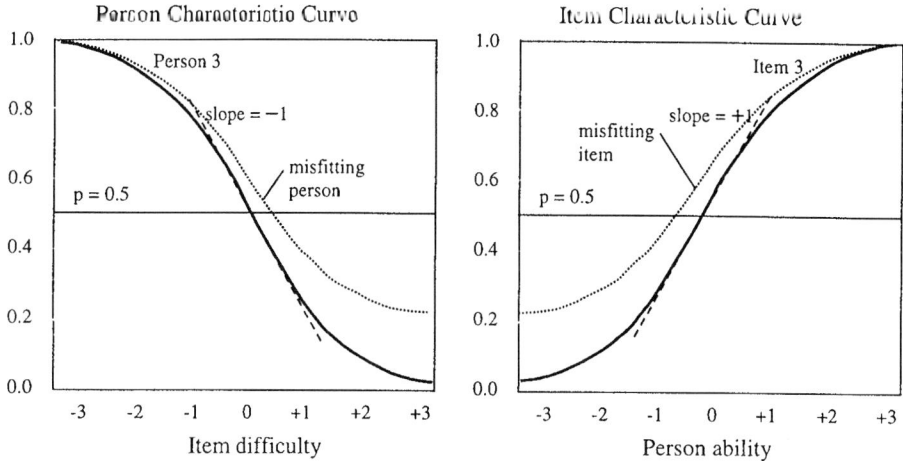

Figure 2 Item and Person characteristic curves

$$(\beta_n - \delta_i - \rho_k - \tau_l - \gamma_m) = \ln \frac{(p_{niklm})}{(1 - p_{niklm})}$$

where n, i, k, l and m are subscripts for the numbers of persons, items, raters, thresholds and stages. (see *Rasch Measurement Models; Partial Credit Model; Rating Scale Analysis; Measurement of Developmental Levels; Measurement of Judgments*).

In addition, if items differ substantially in their discrimination, that is if the item characteristic curves are not of similar gradients when $p = 0.5$, then a discrimination parameter can be introduced into the logistic function item response model to provide for the differences in discriminating powers between items:

$$p(x_{ni} = 1) = \frac{e^{1.7\alpha_i(\beta_n - \delta_i)}}{1 + e^{1.7\alpha_i(\beta_n - \delta_i)}}$$

where α_i is the item discrimination index for item i and 1.7 is a scaling constant to provide equivalence with the normal ogive model (Birnbaum, 1968).

This model is referred to as the two-parameter logistic model. It should be noted that in the calculation of probabilities and scores, a person's response to a particular item is in effect weighted by the discriminating power of the item across all persons in the group under survey. As a consequence the scale formed through the use of this model is no longer independent of the persons and items being surveyed, and the idea of measurement is destroyed beyond the particular situation under consideration.

With multiple choice test items, it is commonly contended that the guessing of responses to an item is a characteristic of the item, and under these circumstances a third parameter can be added to the logistic model:

$$p(x_{ni}=1) = \phi_i + (1-\phi_i)\frac{e^{1.7\alpha_i(\beta_n-\delta_i)}}{1+e^{1.7\alpha_i(\beta_n-\delta_i)}}$$

where ϕ_i is the guessing parameter for item i (see *Item Response Theory*).

The introduction of this third parameter further destroys the idea of measurement in the testing situation. Moreover, very large samples are required for estimation and in many situations convergence is not achieved without arbitrary adjustments. Furthermore, it may be argued that guessing is a characteristic of a person and not necessarily of an item (see *Guessing in Multiple Choice Tests*).

7   Dimensionality and the idea of a latent trait

The key idea on which item response theory is based, is that there is a single underlying latent trait that is being measured, and as a consequence this new approach has been known as latent trait theory to emphasize this idea. The measurement of a latent trait would appear to be readily accepted by those who wish to investigate psychological abilities, but is often challenged when used to assess performance on achievement tests or to assess attitudes, on the grounds that rather more than a single latent trait is involved when achievement and attitudes are being measured. Nevertheless, these critics of latent trait theory, are commonly willing to calculate a total score or a profile of subscores on achievement tests and attitude scales. Moreover, the existence of an underlying latent trait is acknowledged in classical test theory in the concept of "true score".

In the clarification of these issues which arise both in classical test theory and item response theory the construct of unidimensionality was introduced to account for the covariation among responses to a set of test or attitude scale items. Bejar (1983, p. 31) has pointed out that

> undimensionality does not imply performance on the items is due to a single psychological process. In fact, a variety of psychological processes are involved in the act of responding to a set of items. However, as long as they function in unison – that is the performance on each item is affected by the same processes and in the same form – undimensionality will hold.

It is not an assumption of either classical test theory or item response theory that unidimensionality is present, but rather a requirement that the condition of uni-dimensionality must hold before a total test score is calculated under classical test theory, or a scaled score is estimated using items response theory. Consequently, it is necessary to check whether test items satisfy this requirement of unidimensionality before estimation can proceed. In classical test theory, item analysis procedures are employed, and a reliability index is calculated in order to support the meaningfulness of a total test score. Furthermore, when uncertainty exists as to whether a single underlying dimension is present in investigations that employ classical test theory, it is necessary to undertake a principal components analysis or a confirmatory factor analysis to examine the factor structure of the set of test or attitude scale items, prior to combining scores on these items to obtain one or more total scores.

Through the use of item response theory and in particular the Rasch model, an attempt is made to improve the measurement properties of total scores by developing an interval scale that is independent of both the items and the persons who have responded to a particular test or attitude scale. This involves the construction of a more permanent scale of achievement or attitude that demands more stringent tests of unidimensionality. It would seem desirable that as a first step in the construction of a scale of measurement confirmatory factor analysis should be employed to examine the covariation among responses to a set of test or attitude scale items. If support is gained from confirmatory factor analysis for the presence of a single latent trait and for unidimensionality to hold then it would seem appropriate to proceed with more stringent tests using the particular item response theory model considered most appropriate. Bentler and Houck (1996) have pointed out that the use of confirmatory factor analysis would support the use of the two-parameter model referred to above, in which the items differed in their discriminating powers. However, as Wright (1996) contends, Rasch scaling procedures are employed in the calibration of dichotomously and polychotomously scored items, for which the metric between item score values is crude in order to improve the measurement properties of the scaled scores. Under these circumstances it would seem inappropriate to test for unidimensionality using crude confirmatory factor analysis procedures that were equivalent to those employed in classical test theory. Nevertheless, Wright does not provide a test of the structure of responses to a set of items to determine whether unidimensionality could be considered to exist in the set of items.

The dimensional structure that is latent in the responses to a set of items is largely a question of degree of agreement to the requirement for unidimensionality that arises from theory. Consequently tests of fit must be applied to establish whether or not particular items could be considered to meet the conditions demanded by the Rasch model. Those engaged in research in which classical test theory is employed, should continue to examine their data using principal components and confirmatory factor analysis in order to investigate the factor structure of their data. Those seeking to improve the measurement of the abilities or performance of persons using their responses to a set of items, must also establish appropriate procedures for examining the dimensionality of the observed responses of persons that have the characteristics of high fidelity and adequate bandwidth in order to represent in a meaningful and useful way the latent traits that are hypothesized to lie beneath the observed responses. These issues must be expected to result in the development of multidimensional item response models and the specification of the conditions under which such models should be used. This is an area in which advances are likely to occur in the future (see *Latent Trait Measurement Models*). However, before a decision is made to employ a unidimensional analysis or a multidimensional analysis evidence must be obtained as to whether a unidimensional or multidimensional structure best fits the data and is consistent with established theory. An examination of this problem would seem to be urgently needed.

8 Item and person fit

The test for unidimensionality that is widely employed in the use of the Rasch model is the degree of fit of the responses of both the persons and items to the theoretical person and item characteristic curves and the logistic model employed. In Figure 1, dotted lines show the observed person and item characteristic curves that must be tested for fit with respect to the theoretical curves, that are indicated by the continuous lines. The tests of fit assume that the observed probabilities of response are normally distributed and deviate from the expected or theoretical curve by amounts that can be summed and tested. The theoretical curve is specified by the logistic function which with large enough sample sizes approximates to the normal distribution function, so that the significance tests that apply to the normal distribution function can be safely employed.

Since x_{ni}, the observed probability of the response of person n to item i, is normally distributed it has an expected value $\xi(x_{ni}) = p_{ni}$ and variance $Var(x_{ni}) = w_{ni} = p_{ni}(1 - p_{ni})$, i.e.

$$x_{ni} \sim N(p_{ni}, w_{ni})$$

These corresponding standardized residuals are normally distributed with a mean of zero and unit variance, i.e.

$$z_{ni} = \frac{x_{ni} - p_{ni}}{\sqrt{w_{ni}}} \sim N(0, 1)$$

The squared residuals have an expected value of 1 with a chi-square distribution and one degree of freedom, i.e.

$$z^2_{ni} = \frac{(x_{ni} - p_{ni})^2}{w_{ni}} \sim \chi^2_1 \tag{3}$$

This expression in Equation (3) for the standardized residuals can be summed across persons for item i to obtained the *item outfit* or *unweighted* statistic

$$\sum_{n=1}^{N} z^2_{ni} = \sum_{n=1}^{N} \frac{(x_{ni} - p_{ni})^2}{w_{ni}} \sim \chi^2_{N-1}$$

This sum has a chi-square distribution and $N - 1$ degrees of freedom and an expected value of $N - 1$.

The *item outfit* or *unweighted mean square* statistic is the mean of this sum, obtained by dividing by N. The statistic has an expected value of 1 and for large values of N, is a chi-square distribution with one degree of freedom:

$$\frac{\sum\limits_{n=1}^{N} z_{ni}^2}{N} = \frac{\sum\limits_{n=1}^{N} \dfrac{(x_{ni} - p_{ni})^2}{w_{ni}}}{N} \sim \chi_1^2$$

Likewise the expression in Equation (3) can be summed across items for person n to obtain the *person outfit* or *unweighted* statistic.

$$\sum\limits_{i=1}^{L} z_{ni}^2 = \sum\limits_{i=1}^{L} \frac{(x_{ni} - p_{ni})}{w_{ni}} \sim \chi_{L-1}^2$$

This sum has a chi-square distribution with $L - 1$ degrees of freedom and an expected value of $L - 1$.

The *person outfit* or *unweighted mean square* statistic is the mean of this sum, obtained by dividing by L with an expected value of 1 for large values of L as a chi-square distribution with one degree of freedom, i.e.

$$\frac{\sum\limits_{i=1}^{L} z_{ni}^2}{L} = \frac{\sum\limits_{i=1}^{L} \dfrac{(x_{ni} - p_{ni})^2}{w_{ni}}}{L} \sim \chi_1^2$$

Since the expected value of the positive slope of the item characteristic curve for $p(x_{ni}) = 0.5$ is unity when $\beta_n = 0$ the *item outfit* or *unweighted mean square* may be considered as an index of the (slope)^2 with an expected value of $+1$.

Moreover, since the expected value of the negative slope of the person characteristic curve for $p(x_{ni}) = 0.5$ is -1 when $\delta_i = 0$ the *person outfit* or *unweighted mean square* may be considered an index of the (slope)^2 with a value of $+1$.

Thus the *outfit* or *unweighted* statistic for items is summed over persons which are assumed to be independent estimates. The statistic examines the residuals for an item across persons. The *outfit* or *unweighted* statistic for persons is summed over items which are assumed to be independent estimates, and it examines the residuals for a person across items as a consequence, for example, of carelessness on easy items, and guessing on more difficult items. Unfortunately, items are commonly clustered into testlets and persons are clustered into school and classroom groups, and as a consequence the division by N and L to obtain the mean chi-square value does not always provide an appropriate statistic for significance testing. However, the *item outfit* or *unweighted* and the *person outfit* or *unweighted mean square* are statistics whose magnitude has some meaning, with expected values of unity.

The property of particular interest is the slope of the item characteristic curves at their steepest or central point, normally where $p(x_{ni}) = 0.5$. An estimate of the *item infit* or *weighted mean square* statistic at that point is obtained by weighting up the residuals near the central point relative to the residuals at the extremes which are weighted down by using the expression for the variance as the weighting variable. Thus residuals are weighted by

$$w_{ni} = p_{ni}(1 - p_{ni})$$

to obtain the *infit* or *weighted mean square* statistics.

The *item infit* or *weighted mean square* statistic is given by

$$\frac{\displaystyle\sum_{n=1}^{N} \frac{w_{ni}(x_{ni} - p_{ni})^2}{w_{ni}}}{N}$$

$$= \frac{\displaystyle\sum_{n=1}^{N} (x_{ni} - p_{ni})^2}{N} \sim \chi_1^2$$

The *item infit* or *weighted mean square* statistic obtained by dividing by N has a chi-square distribution with one degree of freedom and an expected value of 1.

It should be noted that the chi-square statistics are not normally distributed, but they can be transformed and standardised to yield a *t*-distribution which is sometimes considered to be easier to interpret with values in excess of 2 being considered statistically significant at the five per cent level. Nevertheless, the *t*-expression is influenced by sample sizes and does not provide a very meaningful statistic for large cluster samples of persons and small cluster samples of items.

9 The examination of item fit statistics

It is customary for items to be considered to fit the Rasch model if they have *item infit* or *weighted mean square* statistics with the range 0.77 to 1.30 (Adams and Khoo, 1993), although many researchers would prefer to use a more restricted range from 0.83 to 1.20.

For small samples and short tests, a correction should be applied to the values of the estimates of the *infit* (*weigted*) and *outfit* (*outweighted*) *mean square item* statistic using correction factors of $(L/L - 1)$ and $(N/N - 1)$ to allow for bias.

Andrich (1988) has pointed out that under normal procedures of inquiry in social science and behavioral research the hypothesized model is tested for fit against the data. However, in the procedures described above consideration is always given to excluding persons whose behavior is erratic and items that have been poorly constructed or that operate over a very narrow ability range thereby providing redundant or limited evidence for scale calibration. It is important, however, to examine the manner in which an item performs using classical item analysis procedures, as well as the structure of the item or task, before excluding the item from further analysis. Likewise the response behavior of individual persons and groups of persons can be informative both from an educational and analytical standpoint.

10   The examination of person fit statistics

In the examination of the person statistics for fit to the Rasch model, the outfit or unweighted mean square statistic is sometimes considered to provide more useful information than the infit or weighted mean square statistic does, since the performances of a person on both the easier and the harder items are taken into equal consideration. Moreover, any marked difference between the calculated values for the outfit and infit statistics is highly informative since it indicates a tendency for a different pattern of responding to easier or harder items when compared to items at the center of the scale. Consequently the item response pattern of those persons who exhibit large outfit or unweighted mean square values should always be carefully examined. Where erratic behavior is detected, there would be a strong case for the exclusion of those persons from the analyses for the calibration of the items on the Rasch scale. In particular circumstances consideration should be given to not estimating a score for such persons. These circumstances would include the detection of extensive random guessing behavior or omission of a large proportion of the test items, or responding only to items involving a particular content area. It should be noted that the existence of few items on a scale commonly results in all persons tending to fit the scale. With a large number of items the proportion of misfitting persons tends to increase.

It should also be noted that items which no person answered correctly or which all persons answered correctly do not provide any data that could be used in scale calibration. Likewise persons who obtain a perfect score or who answer no items correctly provide no data that could be used in the examination of the fit of items to the Rasch scale. These items and persons are then automatically excluded from the calibration of the scale. However, those persons with a perfect or zero raw score need to have an appropriate scale score calculated for them.

11   Information

Item response theory through the use of the logistic function not only permits the estimation of the item and person parameters but also permits consideration of how precisely each of the parameters is estimated. The term *information* was first introduced by Fisher (1922) to indicate the precision of estimation.

The slope of the item characteristic curve at any point along the curve indicates the accuracy with which the item differentiates between adjacent ability or performance levels. Mathematically the slope of the line and tangent at each point on the curve is given by the first derivative of the logistic function on which the item characteristic curve is based. Thus information is given by:

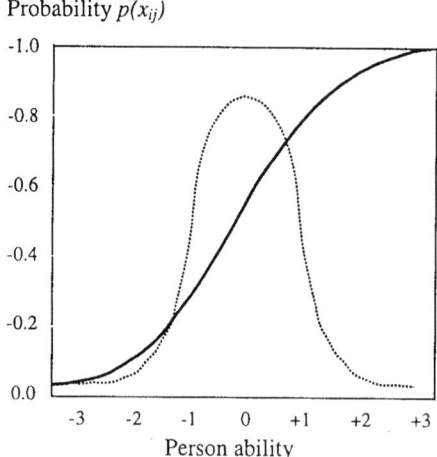

Figure 3 Graph of information and item characteristic curve

$$I(\beta_n) = \frac{(\text{slope})^2}{\text{conditional variance}}$$

Figure 3 shows the graph of the information plotted (as a dotted curve) against person ability for the item characteristic curve (as a continuous curve) for Person 3 ($\beta_3 = 0$) and Item 3 ($\delta_3 = 0$) considered in Table 1 and Figure 1.

If the item characteristic curve of an item is flat with its *infit* or *weighted mean square* index being substantially greater than unity (e.g. > 1.30), then the information value is also reduced in magnitude as indicated by a platokurtic curve. Thus the item provides little information with which to differentiate between persons of adjacent levels of ability or performance.

If the item characteristic curve of an item is steep with its *infit* or *weighted mean square* index being substantially less than unity (e.g. < 0.77) then the information provided is greatly increased in magnitude, but only within a narrow ability range, as indicated by a leptokurtic curve (i.e. it is narrower at the peak). While the item discriminates sharply between persons in adjacent ability or performance ranges, it does so only for a narrow band of abilities or performances.

These comments illustrate through the use of the recorded music analogy (Cronbach, 1960) the necessary compromise between bandwidth and fidelity. The flat information curve indicates that an item is unable to distinguish sharply between adjacent ability levels, but the item spreads across a wide ability or performance range. Such an item could be considered to have a large bandwidth but a low fidelity. A peaked information curve indicates that an item operates effectively in a narrow ability range, with a huge fidelity within that range. However, the item has a very narrow bandwidth and contributes little to the precision of the estimates outside that very narrow range.

It is evident that a balance must be achieved between the bandwidth and fidelity of an item, so that each item provides an optimal degree of information. Moreover, the items included in a test must be spread across the person ability continuum in order to achieve the required range of bandwidth with as high a level of fidelity as possible. The items that would achieve these ends are those that have item characteristic curves that are parallel to the standard item characteristic curve.

Furthermore, conformity to the standard item characteristic curve provides evidence of the unidimensionality of the set of items. The exclusion of items that discriminate too sharply or items that fail to discriminate adequately across the ability or performance range, can be justified on the grounds that such items fail to satisfy a requirement for unidimensionality since these items do not operate consistently with respect to the other items in the scale. Nevertheless, the idea of unidimensionality also involves the items or tasks to which a person is required to respond conceptually measuring the same trait or characteristic. As a consequence a test for unidimensionality would seem to involve either a comparison between alternative clusterings of items or tasks, in order to make a comparative judgment of fit or the setting of a standard for degree of fit that items or tasks must satisfy for the requirement of unidimensionality to hold. Research into the use of multidimensional item response theory, must of necessity, develop empirical tests for both the unidimensionality and the multidimensionality of a data set that extends beyond the conceptual classification of the items and tasks.

12   The estimation of item and person parameters

In the estimation of the parameters for items and persons, the first necessary step is to locate the origin of the scale. The origin (0) of the scale is usually chosen as the mean of the item difficulty parameters, because the items are commonly explicitly selected for inclusion in a test, whereas the persons are commonly randomly chosen. As a consequence the items provide a more readily interpretable fixed point for the scale. However, the persons and the items must of necessity satisfy the requirement of conforming to Equation (2) and those items that all persons or no persons answered correctly provide no information, as do these persons who answered all items or no items correctly.

Conditional probabilities permit the elimination of the person parameters from the item parameters, since these parameters are simple sufficient statistics for both persons and items in the Rasch model (see *Sufficient Statistics in Educational Measurement*). A sufficient statistic for a parameter is one that contains all the modeled information in the data with respect to that parameter. Thus the score of a person, or the ability of a person given the difficulty levels of the items that the person responded to is simply a function of the proportion of items that the person answered correctly. Likewise the difficulty level of an item, given the average difficulty of the items being set at zero, is simply a function of the proportional of persons who responded correctly to the item.

A table for the calculation of person abilities is contructed for the proportions of correct responses as if all persons responded to all items. However, if a person omitted particular items, that person's ability can be estimated with a scale zero set at the mean difficulty level for the items to which the person responded, and a score calculated on the proportion of correctly answered items to the total number of items attempted by that person together with an adjustment made for the scale difference between that person's zero point and the scale zero.

The error associated with an estimated scale score can be derived from the combined information functions of the items to which a person responded. If this combined information function includes all items in a test, it is referred to as the test information function. Where a person responded to many items close to that person's level of ability or performance, the combined information function peaks close to the person's estimated ability level, and the error of measurement which is derived from the combined information function is smaller than it would be if the person responded to items that were remote from the estimated ability level. However, in the use of the Rasch model, it is common to employ the error of a proportion to estimate the error associated with a scale score using a simple formula.

If the two or three parameter logistic model is employed in the calibration of the items in a scale, then a maximum likelihood estimation procedure or the empirical Bayes method is used in the calibration of the scale score for a person. These scoring procedures optimally weight the person's responses, with allowance made for the discrimination of the item and a guessing factor where appropriate. However, such scores are provided on a scale that is not independent of the items and persons used in calibration.

It should be noted that each estimated scale value has its own standard error of measurement, that is calculated from the combined or test information function for the items responded to by the person taking the test. The statistical information provided by an item is the reciprocal of the precision with which a parameter can be estimated. Consequently to estimate the error of measurement it is necessary to take the reciprocal of the square root of the information available for the items responded to by a person at a particular ability level.

Only with the Rasch or one parameter logistic model is true measurement achieved on a scale that is independent of the items and persons involved in calibration. However, in the estimation of person scores and the error of measurement of these scores, the empirical Bayes procedure could be employed.

In conclusion, the calculation of scaled scores using the item response theory requires two separate operations which are best considered not only as conceptually different, but are also best carried out as two different stages. The first stage involves the *calibration* of the items or tasks that form the scale. Decisions must be made with respect to which items and persons should be included in calibration, since it is not necessary for all items and persons to be involved. Clearly, those items and persons providing no information should be excluded. Likewise those items and persons that do not satisfy the requirements of unidimensionality, which is interpreted as conforming to the Rasch model or other item response model that is employed must be removed from consideration in calibration. The parameter estimates obtained from calibration are used to define the scale, and set the fixed or zero point of the scale.

The second stage involves *scoring*. Decisions must also be made with respect to which items should be included in the scoring operation, and for which persons a score could be meaningfully estimated. A subsequent step involves the calculation of the errors of measurement for each score value and each item difficulty level.

13 Missing responses

Major problems arise in the operations of calibration and scoring from the presence of missing responses. In any testing program, missing data occurs as a result of a student failing, by accident or by intention to provide a response to an item in the body of a test. These failures to respond are referred to as "omitted responses". Failure to respond to items also occurs at or near the end of a test, as a consequence of a person running out of time to complete a test. These omissions are referred to as "not-reached responses". Neither occur at random. Different decisions need to be made with respect to the treatment of omitted and not-reached responses in both calibration and scoring. Furthermore, the decisions made can differ under different testing conditions. What is clear is that if too much missing data are present in a data set, suggested by some to be in excess of 15 per cent, then substantial bias is introduced whatever the steps taken to deal with the missing data. The decisions to be made with respect to the treatment of missing data must be carefully considered at all stages of planning and the conduct of an investigation. Many otherwise well planned research studies and testing programs have foundered on a failure to plan adequately for the processing of data where responses are missing.

14 Equating

The problems of equating across different instruments, different samples and different occasions also require careful consideration. The advantages of using the Rasch measurement model is that the procedures involved in equating are greatly simplified. The procedures employed in equating under the Rasch measurement model involve: (a) the common item difference method, (b) the anchor item method, and (c) the concurrent equating method. Moreover, since items and persons have a fully reciprocal relationship in the logistic function that is employed in the Rasch measurement, items and persons could be used interchangeably. However, items are commonly used rather than persons in equating simply because there is more control over items than over persons.

The most powerful method of equating is emerging as the concurrent equating procedure (Morrison and Fitzpatrick, 1992; Mohandas, 1996), because it is now possible to merge relatively large data files with a large number of persons and a large number of items on to a common file so that the calibration of two or more data sets can be carried out simultaneously. However, what would appear to be lacking is the calculation of appropriate estimates of error arising from this equating process. Work is required to obtain such estimates now that this equating procedure is being used increasingly in practice and since the method of estimating equating error greatly influences the number of common items and common persons employed in equating.

15 Establishing the fixed point

The interval scale provided by the Rasch model has a natural metric, the logit, but requires that a single fixed point should be identified in order to locate the scale along the number line. In so far as calibration of the scale is customarily carried out using a well-constructed test from which items that do not fit the scale have been eliminated, the mean difficulty level of the items in the test would seem to provide an appropriate fixed point for the scale. While the mean ability level of the sample of persons employed in calibration could be used for establishing the fixed point of the scale, in general, such a sample would not seem to provide a well-identified group with which to locate the scale along the number line. It is clear that the items in an instrument have greater permanence than the persons used in calibration. Once the fixed point of a scale is determined using a standard instrument, other instruments measuring the same unidimensional construct can be located on that scale using equating procedures.

16 Estimation processes

In the initial development of Rasch scaling procedures approximate methods were employed in order to reduce the computation involved. As an example in the article on *Individualized Testing in the Classroom* the PROX algorithm (Cohen, 1979) is presented in which an approximation to the normal distribution is employed. The BIGSCALE (Linacre, 1989) and QUEST (Adams and Khoo, 1993) programs use the UCON algorithm, which involves an unconditional maximum likelihood estimation procedure (see *Rasch Measurement Models*). The RUMM program (Andrich *et al.*, 1996) employs a paired comparison algorithm, which uses a fully conditional maximum likelihood estimation procedure (Choppin, 1985). The CONQUEST (Wu, Adams and Wilson, 1996) program uses marginal maximum likelihood estimation procedures (Wilson and Adams, 1993). The use of these different procedures has been made possible through the development of more powerful computers for the processing of data.

Two aspects of the estimation procedure would seem to be of importance. First, it is necessary to consider the minimum number of items and persons necessary in order to achieve stable estimates. It is recommended that with BIGSCALE and QUEST that 50 persons should be employed in estimation for each response category. Thus for dichotomous test items at least 100 persons are required and for trichotomous attitude scales at least 150 persons are needed. In addition, it would seem likely that at least 10 items are required to obtain stable estimates of the parameters of the item characteristic curves used in calibration. However, RUMM, using the paired comparison algorithm obtains stable estimates with fewer persons and items.

Secondly, it is necessary to consider the efficiency of the estimation procedure, which while dependent on the numbers of items and persons employed would also appear to be dependent on the algorithms used in estimation. The CONQUEST (Wu, Adams and Wilson, 1996) program undertakes estimation with smaller standard errors of both the peron and item estimates.

Work is urgently needed on the characteristics of the different computer programs that are now available for the estimation of Rasch scaled scores, together with information on precision and efficiency, robustness with misfitting items and persons, and accuracy over the scale range for the different programs.

References

Adams, R. J. and Khoo, S. T. 1993. *QUEST: The Interactive Test Analysis System*. ACER, Melbourne.

Andrich, D. 1988. A general form of Rasch's extended logistic model for partial credit scoring. *Applied Measurement in Education* **1**(4), 363–78.

Andrich, D., Luo, G. and Sheridan, B. 1996. *RUMM. Rasch Unidimensional Measurement Models*. Murdoch University, Perth.

Bejar, I. I. 1983. *Achievement Testing: Recent Advances*. Sage Publications, Beverly Hills, California.

Bentler, P. M. and Houck, E. L. 1996. Structural equation modeling multiple sclerosis disease status. *The International Test Commission Newsletter* **6**(1), 11–13.

Birnbaum, A. 1968. Some latent trait models and their use in inferring an examinee's ability. In: Lord, F. M. and Novick, M. R. (eds.) 1968. *Statistical Theories of Mental Test Scores*. Addison-Wesley, Reading, Massachusetts.

Choppin, B. 1985. A fully conditional estimation procedure for Rasch model parameters. *Evaluation in Education* **9**, 29–42.

Cohen, L. 1979. Approximate expressions for parameter estimates in the Rasch model. *Br. J. Math. S.* **32**(1), 113–20.

Cronbach, L. 1960. *Essentials of Psychological Testing*. Harper and Row, New York.

Ferguson, A. G. 1942. Item selection by the constant process. *Psychometrika* **7**, 19–29.

Fisher, R. A. 1922. On the mathematical foundations of theoretical statistics. *Phil Trans. Royal Soc. A* **222**, 309–68.

Guttbaum, L. 1944. A basis for scaling qualitative data. *American Sociological Review* **9**, 139–50.

Lawley, D. N. 1943. On problems connected with item selection and test construction. *Proc. Royal Soc. Edin.* **61**, 273–87.

Linacre, J. M. 1989. *BIGSCALE: A Rasch Measurement. Computer Program*. MESA Press, Chicago, Illinois.

Lord, F. M. and Novick, M. R. 1968. *Statistical Theories of Mental Test Scores*. Addison-Wesley, Reading, Massachusetts.

Mohandas, R. 1996. Test Equating, Problems and Solutions. Unpublished Med Thesis, Flinders University of South Australia.

Morrison, C. A. and Fitzpatrick, S. J. 1992. *Direct and Indirect Equating: A Comparison of Four Methods Using the Rasch Model*. Measurement and Evaluation Center, University of Texas at Austin, Austin, Texas.

Rasch, G. 1960 (reprinted 1980). *Probabilistic Models for Some Intelligence and Attainment Tests*. University of Chicago Press, Chicago, Illinois.

Thurstone, L. L. 1927. Psychological Analysis. *American Journal of Psychology* **38**, 368–89.

Wilson, M. and Adams, R. J. 1993. Marginal maximum likelihood estimation for the ordered partition model. *J. Ed. Stat.* **18**(1), 69–90.

Wright, B. D. 1996. Comparing Rasch measurement with factor analysis. *Structural Equation Modeling* **3**(1), 3–24.

Wu, M., Adams, R. J. and Wilson, M. 1996. *CONQUEST.* ACER, Melbourne.

3 Latent Trait Measurement Models

H. Swaminathan

Latent trait theory, in general terms, is concerned with the identification of traits such as aptitudes, interests, cognitive abilities, and personality variables that give rise to observed behavior. These latent traits are unobserved but it is assumed that the values the subject has on them influences, or predicts, in a stochastic sense, the observed response(s) of the subject. The goal of latent trait measurement theory is to infer, based on the observed responses, the subject's values on these latent traits. In order to accomplish this goal, a statistical model which specifies the relationship between the observed responses and the set of latent traits is postulated. Once the statistical model is specified, the values the subject has on the latent traits is estimated using standard statistical procedures.

The statistical model that is used to specify the relationship between the observed response and the latent traits could take many forms; it could be linear, curvilinear, or nonlinear. The choice of model depends upon the nature of the observed response, the nature of the latent trait, and finally, the mathematical form that is deemed appropriate. The model may be taken as unidimensional if only one trait is believed to influence the observed responses, or multidimensional if more than one trait is invoked.

1 Linear latent trait measurement models

While the term "latent trait" appears to have been first introduced by Paul Lazarsfeld in 1950, the concept goes back to Spearman in the early years of the twentieth century. Spearman (1904) postulated that the observed score y_{ji} of examinee i on test j is related to the examinee's unobservable or latent ability, θ_i, according to the linear model

$$y_{ji} = \mu_j + f_j \theta_i + e_{ji} \tag{1}$$

where f_j is, in factor analysis terms, the "factor loading", and e_{ji} is the error or "unique score". The term μ_j is an additive constant which has no effect on the correlation between the observed score and the trait. It is included here for the sake of completeness.

The unifactor theory of Spearman was rejected by Thurstone who postulated that the correlations among scores on a battery of tests can be more reasonably explained by a set of traits, which he termed "common factors". The factor model postulated by Thurstone take the form

$$y_{ji} = \mu_j + f_{j1}\theta_{1i} + f_{j2}\theta_{2i} + \ldots + f_{jk}\theta_{ki} + e_{ji} \qquad (2)$$

where $f_{j1}, f_{j2}, \ldots, f_{jk}$ are the factor loadings associated with the k latent traits $\theta_1, \theta_2, \ldots, \theta_k$. As with the unifactor model, the purpose of the multifactor model was to provide an explanation of the correlations among the observed test scores.

The primary objective in measurement is to determine, through observations, the scores an examinee has on a set of latent traits that give rise to the observations. While the factor models proposed by Spearman and Thurstone were designed to provide an exaplanation of the observed correlations among the test scores in terms of the underlying latent trait, they also serve as measurement models through which the values the examinee has on the latent traits can be determined. The Spearman model can be thought of as a unidimensional measurement model and the Thurstone model as its multidimensional extension.

In principle, the factor models given by Equations (1) or (2) can serve as latent trait measurement models and have been used as such, albeit in a limited sense, in personality assessment and vocational classifications. In achievement and aptitude testing factor analytic methods have been used primarily in test construction, particularly for item selection and validity assessment, and not as a framework for measurement.

Classical test theory (Lord and Novick, 1968), on the other hand, has provided the framework for measurement during the past few decades. The foundation of classical test theory lies in the decomposition of the observed score in terms of true score and error score that is

$$y_{ji} = \tau_{ij} + e_{ij} \qquad (3)$$

where τ_{ij} is the true score of examinee i on test j and e_{ij} is the error score. In comparing Equation (3) with the Spearman model given in Equation (1), it can be seen that τ_{ij} is further decomposed. While the classical test theory decomposition is a tautology (Lord and Novick, 1968), the decomposition given by the factor model by Equation (1) is falsifiable, i.e., the Spearman model is a statistical model which may or may not fit the data.

The simplicity of classical test theory and the fact that the classical test theory "model" is not subject to model-data fit, makes it appealing. However, it has serious drawbacks in that the indices that characterize an item, that is, item difficulty and item discrimination, and that which characterize the test, that is, reliability, are not invariant across examinee subpopulations. Similarly, the score that is assigned to an examinee is dependent on the particular set of items administered to the examinee; consequently, the concept of parallel test forms is needed for comparing performance of examinees and for the definition of the reliability of the test. In addition, the standard error of measurement associated with the score assigned to the examinee is assumed to be the same for all examinees. As a result, the standard error of measurement is an average value across the score range of the examinees, and in one sense, not appropriate for any examinee. Finally, classical test theory does not provide a framework for assembling items in a test to meet a certain objective, e.g., assembling a set of items to yield a desired precision of measurement at a specified trait value (for more detail and further discussion on these issues see Lord, 1980; Hambleton and Swaminathan, 1985) (see *Item Response Theory*).

The need for a better framework for measurement than that provided by classical test theory has prompted measurement theorists to consider model-based measurement procedures. The simplest model for measurement is the Spearman model, given by Equation (1). Since in this model the observed score of an examinee on a test is a function of the characteristic of the test and the trait level of the examinee, this model provides a framework for taking into account the interaction between the trait level of the examinee and the characteristic of the item. While this linear latent trait model is appropriate for relating test scores to underlying traits, it is not appropriate for relating an examinee's trait to his/her performance on a dichotomously scored item. Equation (1) represents the regression of the score y_{ji} on the latent trait θ_i and hence is the expected value of y_{ji} conditional on the value of θ_i. When y_{ji} is dichotomous, this expected value is the probability that $y_{ji} = 1$. Since the probability is bounded by zero and unity, the regression curve cannot be linear. Thus when items are scored dichotomously, non-linear models that relate the probability of a correct response to the underlying trait are required.

2 Nonlinear latent trait measurement models

2.1 *Dichotomous item response models*

The problem of predicting a dichotomous dependent variable from an observed regressor variable was first addressed by Finney in the early 1930s in the context of bioassay. Finney chose the cumulative normal probability function (the area under the normal curve) to model the relationship between the probability of "success" and an observed regressor. This normal ogive model, or the probit model was extended by Lawley (1943) to the situation where the regressor was a latent variable. Lord (1952, 1953), in extending the work of Lawley, formalized latent trait theory, by introducing the "assumption of local independence".

The assumption of local independence, according to Lord and Novick (1968), ". . . is the foundation of latent trait theory". The assumption of local independence or conditional independence states that once all the latent traits that contribute to performance on a set of items are determined, the responses to the items are statistically independent. This assumption is similar to that found in factor analysis; when the factors that contribute to the correlations among observed variables are partialed out, the observed variables will cease to be correlated. While the assumption of local independence may seem obvious or trivial when stated in this manner, the implications are profound. The assumption of local independence is equivalent to the assumption that all the traits that are relevant are included in the latent trait model; that is, the complete latent space is specified.

The major issue with the assumption of local independence revolves around the assumed dimensionality of the latent space. if the assumed dimensionality is correct, then local independence holds. However, for convenience and for mathematical reasons a unidimensional latent trait model is usually assumed. If the complete latent space is not unidimensional, the assumption of local independence will not hold. Up until the mid-1990s no foolproof method existed for examining the dimensionality of the latent space. Factor analysis is an obvious choice for examining the dimensionality of the latent space. However, as noted earlier, factor analysis was considered inadequate to model the relationship between a dichotomous, observed variable and a continuous latent variable. Given this, it does not make much sense to use the model that was deemed inappropriate to check the assumptions underlying a model that replaced it.

Like the factor model, "item response theory" (a term coined by Lord in the late 1970s to replace the more general term, "latent trait theory") is based on the assumption that an examinee's response to an item is a function of the examinee's trait level (also known as ability level or proficiency level) and the characteristics of the item. In contrast to the model in Equation (1), in item response theory it is the probability of a correct response that is modeled. The higher the proficiency level, the higher the probability of responding correctly to the item. This probability is also affected by the characteristics of the item; if the item is very "difficult" relative to the examinee's proficiency level, the probability of a correct response will be low (see *Item Response Theory*).

An item may be characterized by a number of parameters; if the item is characterized by only one parameter, the "difficulty" parameter, the item response model is called a "one-parameter" model. If the item is characterized by the difficulty parameter and the "discrimination" parameter, the item response model is termed a "two-parameter" model. A three-parameter model is one which specifies that examinees with very low proficiency level will have a nonzero probability of responding correctly to the item.

The two-parameter normal ogive model that relates the probability of correct response to an examinee's latent trait and the characteristics of the item is given below:

$$P[y_{ji} = 1 | \theta_i] = \int_{-x}^{a_j(\theta_i - b_j)} e^{-1/2z^2} \, dz \equiv \Phi[a_j(\theta_i - b_j)] \qquad (4)$$

In the expression given above, y_{ji} is the response of examinee i on item j, a_j and b_j are the parameters that characterize the item, and θ_i is the latent trait of examinee i. The function $\Phi(x)$ yields a notational simplification of the normal ogive model.

It is easily seen that in this model, the probability is bounded between zero and one. When $\theta_i = b_j$, the probability of a correct response is 0.5. The higher the value of b_j, the

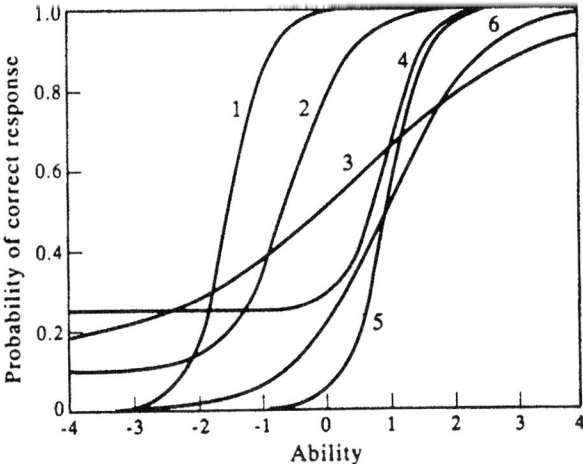

Figure 1 Six item response models

higher the value of θ_i that is needed to have a 50 per cent chance of passing the item. It is in this sense that this parameter is called the "difficulty parameter". Curves 1, 5 and 6 (corresponding to items 1, 5 and 6) in Figure 1, follow two-parameter item response models. By examining the θ corresponding to $P = 0.5$, it can be seen that Item 1 has the lowest b value; Items 5 and 6 have higher but the same b value.

The slope of the curve given by Equation (4) at the point where the probability is 0.5 corresponds to the value of the parameter a_j. The parameter a_j is called the discrimination parameter since as this value increases, the curve rises sharply at the point $\theta_i = b_j$. This implies that examinees with trait values higher than b_j will have a markedly higher probability of responding correctly to the item than examinees who have trait values lower than b_j. The curves corresponding to items 1 and 5 in Figure 1 have the same discrimination parameter value. The curve corresponding to Item 6 rises less sharply than these two curves and hence Item 6 is less discriminating than either Item 1 or 5.

The similarity between the item response model given by Equation (4) and the Spearman model given by Equation (4) is noteworthy. Clearly Equation (4) is the nonlinear analogue of the model given in Equation (1). Thus the item response models presented above are nonlinear factor models. To the extent that the item response function approximates a linear function, the results obtained using item response theory and linear factor model will be in agreement.

Curves corresponding to Items 2, 3 and 4 in Figure 1 have nonzero probability of a correct response as θ_i approaches negative infinity. These curves are given by the equation

$$P[y_{ji} = 1 \mid \theta_i] = c_j + (1 - c_j)\Phi[a_j(\theta_i - b_j)] \tag{5}$$

The parameter c_j corresponds to the lower asymptote; Item 4 has the largest value for the c-parameter; and Item 2 has the lowest c-parameter value of the three items. Obviously, Items 1, 5 and 6 have the lowest possible values for the c-parameter, zero.

2.1.1 Logistic item response model

The normal ogive models are mathematically cumbersome. Berkson (1953) introduced the logistic cumulative distribution function to model the probability of success in the context of bio-assay. Birnbaum (1968) introduced these models in the mental measurement context.

The logistic function is given by the expression

$$P[y_{ji}=1|\theta_i]=c_j+(1-c_j)\frac{e^{1.7a_j(\theta_i-b_j)}}{1+e^{1.7a_j(\theta_i-b_j)}} \tag{6}$$

where the parameters have the same significance as in the normal ogive model. The scaling factor of 1.7 is introduced in the model to make the function as similar as possible to the normal ogive function. In almost all item response theory applications, the logistic model is the model of choice. The two-parameter model is obtained by setting $c_j=0$ in Equation (6).

2.1.2 The invariance property

The item parameter in an item response model are parameters that characterize a curve. The values of these parameters must necessarily be the same along the points on the trait continuum. The parameters are therefore the same across subpopulations of examinees. It should be noted that the parameters obtained using the linear model given by Equation (1) will also be invariant when covariances rather than correlations are analyzed.

The parameter that characterizes an examinee's proficiency level is also invariant across different subsets of items as long as the dimensionality of the latent space remains the same. This invariance property distinguishes item response theory from classical test theory and serves as a powerful tool in test construction and trait assessment.

2.1.3 The Rasch model

When $c_j=0$, and when the item response models do not differ in discrimination, that is the item response functions have the same slope, the one-parameter model results. In Figure 1 Items 1 and 5 have the same discrimination parameter values. These two items taken together follow the one-parameter model. The one-parameter model is usually written as

$$P[y_{ji}=1|\theta_i]=\frac{e^{(\theta_i-b_j)}}{1+e^{(\theta_i-b_j)}} \tag{7}$$

While it is convenient to think of the one-parameter model as a special case of the two- and the three-parameter models, the one-parameter model was developed independently by Rasch (1960) using a totally different approach. Rasch was concerned with the principle of "objective measurement" – he felt that the measurement of an examinee's proficiency level should in no way depend on the items used in the test. Similarly, two items should be comparable without any reference to the group of examinees to whom the item is administered. Using these two criteria he arrived at the one-parameter logistic model, or the Rasch model (see *Rasch Measurement Theory; Rasch Measurement Models*).

In the Rasch model, it can be shown that the number right score contains all the information regarding an examinee's proficiency level, that is, two examinees who have the same number correct score have the same proficiency level. By conditioning the likelihood function (loosely speaking, the joint probability of the observations) on the number right score, it is possible to estimate the parameters that characterize the items without any reference to the examinees. In theory, it is also possible to estimate an examinee's proficiency level independently of the items administered by conditioning the likelihood function on the number of examinees who respond correctly to each item. This procedure, however, cannot be implemented in practice because of the excessive number of combinations that will result. The conditional estimation procedure described above is not possible with the multiparameter models; the invariance property has to be invoked with these models.

2.1.4 Features of the item response theory framework

When the item response model fits the data, the advantages afforded by the theory can be realized. For a complete discussion see Hambleton and Swaminathan (1985), Lord (1980). A few of the advantages are listed below.

(a) Item parameters are invariant across examinee subpopulations.
(b) Trait ability, or proficiency level parameters are independent of the set of items administered to the examinees.
(c) Precision of the trait estimates, or the standard error of measurement, can be determined at each trait level. This obviates the concept of reliability.
(d) The concept of information (inverse of the square of the standard error of measurement) provides a framework for test construction; the information provided by each item can be aggregated. Thus the contribution of each item to the precision of measurement can be assessed and this provides a basis for item selection.
(e) The invariance property of the trait parameters permits the administration of different items to different examinees. Testing can be "adapted" to an individual's trait level. This adaptive testing procedure has been shown to reduce testing time by at least 50 per cent.

2.2 Models for polytomously scored items

The models considered in the above sections are appropriate when the item is scored dichotomously. There are many situations where the items may be scored nondichotomously, that is, polytomously. Ratings on a Likert scale, attitude measurement, and essay scoring are all examples where the items are polytomously scored. In addition in these situations the score categories are ordered. Scoring where the categories are unordered also occur as when the choice of distractors in multiple choice items is considered. Models for these types of scoring can be classified as nominal response models and ordinal response models (see *Rating Scale Analysis*).

2.2.1 The nominal response model

The nominal response model where the response categories are unordered was first considered by Bock (1972). This model can be considered as an extension of the dichotomous model by considering the probability of choosing one category over another. Since the categories are unordered, the probability of choosing a category over the first category may be modeled. If π_j and π_0 are the probabilities of falling in categories j and 0 respectively, ($j = 1, 2, \ldots, k$), then, for an examinee with trait level θ (subscripts for the item and examinee are dropped for notational simplicity),

$$\frac{\pi_j}{\pi_j + \pi_{j-1}} = \frac{e^{a(\theta - b_j)}}{1 + e^{a_j(\theta - b_j)}} \tag{8}$$

where a_j and b_j are the parameters corresponding to the difference between category j and 0. It follows then that

$$\pi_j = \pi_0\, e^{a_j(\theta - b_j)} \tag{9}$$

Since $\pi_0 + \pi_1 + \ldots + \pi_k = 1$, π_0 can be eliminated from the model to yield the probability of failing or choosing category j:

$$\pi_j = \frac{e^{a_j(\theta - b_j)}}{1 + \sum_{j=1}^{k} e^{a_j(\theta - b_j)}} \tag{10}$$

In this model, the examinee's trait value along with the category parameters determine which category the examinee chooses. A Rasch model formulation is obtained by specifying a common a parameter value across the categories and across the items.

2.2.2 The ordinal response model

In the ordinal response model the categories are ordered. In developing the model for this situation, the probability of choosing one category over the "preceding" category must be considered. Thus

$$\frac{\pi_j}{\pi_j + \pi_{j-1}} = \frac{e^{a(\theta - b_j)}}{1 + e^{a(\theta - b_j)}} \tag{11}$$

The important distinction between this model and the model in Equation (10) is that the discrimination parameter is not allowed to vary across the categories since otherwise the order will not be meaningful. The discrimination parameter, however, is permitted to vary across items. It follows that

$$\pi_j = \pi_{j-1}\, e^{\,a(\theta - b_j)} \tag{12}$$

This model yields the probability of choosing category $1, 2, 3, \ldots, k$ in terms of the probability of choosing category 0:

$$\pi_j = \frac{e^{z_j}}{1 + \displaystyle\sum_{r=1}^{k} e^{z_r}},$$

where

$$z_t = \sum_{m=1}^{t} a(\theta - b_m). \tag{13}$$

The model given in Equation (13) is essentially that developed first by Andrich (1978, 1982) in the analysis of rating scales and by Masters (1982) in the context of partial credit scoring. These authors used a Rasch formulation of the model. Andrich (1978), by decomposing the parameter b_j into components specific to the item and to the response category, i.e., $b_j = b + t_j$, was able to provide a more finely tuned analysis of the choice process. The partial credit models of Andrich and Masters, and their generalization have important applications in performance assessment and essay scoring (see *Rating Scale Analysis; Partial Credit Model*).

Samejima (1969) introduced an alternative model, the graded response model, for ordered responses. This model considers the probability that an examinee will respond *at or above* category *j*. In this model the probability that an examinee will choose a particular category is not easily determined. Comparisons between the partial credit and the graded response models are provided by Masters (1982).

3   Multidimensional item response models

The item response models that were in use in the early 1990s were unidimensional models. The assumption that the latent space is unidimensional has been a source of concern for many practitioners and psychometricians. Unlike the factor analysis situation where a multidimensional extension of the Spearman model has been operationalized and developed fully, the multidimensional extension of the item response model is still in its infancy. While multidimensional models have been formulated by several researchers (see

McDonald, 1982; Bock *et al.*, 1985; McKinley and Reckase, 1982; Reckase and McKinley, 1982; Sympson, 1978) the models and procedures are not in operation.

Multidimensional item response models can be classified into compensatory models and noncompensatory models. In the compensatory model an examinee's trait value on one dimension compensates for the trait in another dimension, that is in a mathematics item, an examinee's mathematical knowledge may compensate for his/her lack of language skills. In the noncompensatory model an examinee is required to have a minimum value on all the relevant dimensions in order to respond correctly to the item. The mathematical form of the compensatory model (Reckase and McKinley, 1982) is:

$$P[u=1 \mid \theta_1, \theta_2] = \frac{e^{a_1(\theta_1 - b_1) + a_2(\theta_2 - b_2)}}{1 + e^{a_1(\theta_1 - b_1) + a_2(\theta_2 - b_2)}} \tag{14}$$

where θ_1, θ_2 are the trait values of the examinee on two dimensions, a_1, a_2 are the discrimination parameters on the two dimensions, and b_1, and b_2 are the difficulties on these dimensions. Reckase and McKinley (1983) have provided interpretations of the item parameters. The noncompensatory model was formulated by Sympson (1978). The two-dimensional model is given as a product of two unidimensional item response models, that is,

$$P[u=1 \mid \theta_1, \theta_2] = \frac{e^{a_1(\theta_1 - b_1)}}{1 + e^{a_1(\theta_1 - b_1)}} \times \frac{e^{a_2(\theta_2 - b_2)}}{1 + e^{a_2(\theta_2 - b_2)}} \tag{15}$$

Figures 2a and 2b provide illustrations of the compensatory and noncompensatory two-dimensional item response models. Despite the formulation of these models little is known about their utility and applicability to practical testing problems.

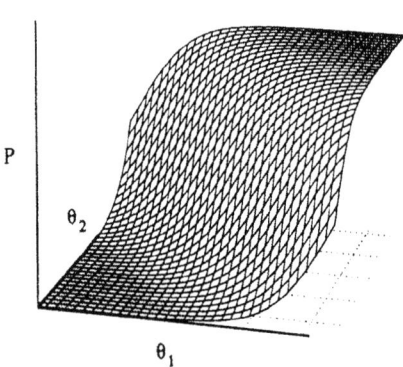

Figure 2a   A compensatory model

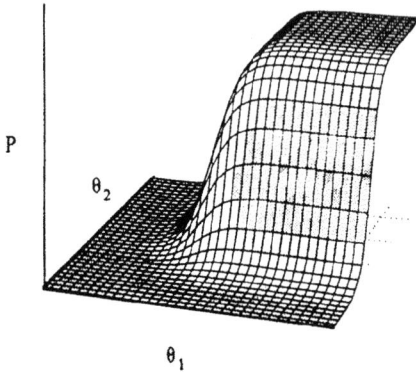

Figure 2b   A noncompensatory model

4 Conclusion

Latent trait measurement theory is a model based measurement theory that provides a general framework for testing and measurement. While linear latent trait models are appropriate when the observed responses are continuous, they are inappropriate when responses to items are dichotomously or polytomously scored. Item response models, which are non-linear latent trait models, are appropriate in the latter situation. Consequently, item response theory has emerged as the modern measurement theory. Procedures based on item response theory are the procedures of choice for test construction, trait estimation, and equating of tests. However, current interest in performance assessment and the possible move away from multiple choice items have resulted in an increased interest in models for nondichotomously scored items. While these models are in place, considerable research is needed before the models can be fully operationalized and used by practitioners. A major area of concern is with the unidimensional nature of the item response models. While research in multidimensional item response models is continuing, little progress has been made to date in the estimation of parameters and in understanding how the models could be applied. Considerable research efforts can be expected in this direction in the late 1990s and beyond (see also: *Item Response Theory*).

References

Andrich, D. 1978. A rating formulation for ordered response categories. *Psychometri.* **43**, 561–73.

Andrich, D. 1982. An extension of the Rasch model for ratings providing both location and dispersion parameters. *Psychometri.* **47**, 103–13.

Berkson, J. 1953. A statistically precise and relatively simple method of estimating the bio-assay with quantal response, based on the logistic function. *J. Am. Stat. Assoc.* **48**, 565–99.

Birnbaum, A. 1968. Some latent trait models and their use in inferring an examinee's ability. In: Lord, F. M., Novick, M. R. 1968. *Statistical Theories of Mental Test Scores*. Addison-Wesley, Reading.

Bock, R. D. 1972. Estimating item parameters and latent ability when responses are scored in two or more nominal categories. *Psychometri.* **37**, 29–51.

Bock, R. D., Gibbons, R. and Muraki, E. 1985. *Full Information Factor Analysis*. MRC Report 85-1. Nat. Opinion Res. Ct. Chicago, Illinois.

Hambleton, R. K. and Swaminathan, H. 1985. *Item Response Theory: Principles and Applications*. Kluwer-Nijhoff, Boston, Massachusetts.

Lawley, D. N. 1943. On problems connected with item selection and test construction. *Proc. Royal Soc. Edin.* **61**, 273–87.

Lord, F. M. 1952. A theory of test scores. *Psychometric Monograph* **7**. University of Chicago Press, Chicago, Illinois.

Lord, F. M. 1953. The application of confidence intervals and maximum likelihood to the estimation of an examinee's ability. *Psychometri.* **18**, 57–75.

Lord, F. M. 1980. *Applications of item response theory to practical testing* Erlbaum, Hillsdale, New Jersey. ERLB.

Lord, F. M. and Novick, M. R. 1968. *Statistical theories of mental test scores*. Addison-Wesley, Reading.

McDonald, R. P. 1982. Linear versus nonlinear models in item response theory. *App. Psych. Meas.* **4**, 379–96.

McKinley, R. L. and Reckase, M. D. 1982. *An extension of the two-parameter logistic model to the multidimensional latent space*. Res. Rep. ONR 83-2. The American College Testing Program, Iowa City, Iowa.

Masters, G. N. 1982. A Rasch model for partial credit scoring. *Psychometri.* **47**, 149–74.

Reckase, M. D. and McKinley, R. L. 1982. The feasibility of multidimensional latent trait model. Paper presented at the meeting of the Am. Psych. Assoc., Washington, DC.

Rasch, G. 1960. *Probabilistic models for some intelligence and attainment tests.* Danish Institute of Educational Research, Copenhagen.

Samejima, F. 1969. Estimation of ability using a response pattern of graded scores. *Psychometric Monograph* Vol. 34, No. 17, Part 2. University of Chicago Press, Chicago, Illinois.

Spearman, C. 1904. General intelligence, objectively determined and measured. *Am. J. Psych.* **15**, 201–33.

Sympson, J. B. 1978. A model for testing with multidimensional items. In: Weiss, D. (ed.) 1978. *Proceedings of the 1977 Computerized Adaptive Testing Conference.* University of Minnesota, Minneapolis, Minnesota.

4 Item Response Theory

M. L. Stocking

Item response theory (IRT) models the relationship between a person's level on the trait being measured by a test and the person's response to a test item or question (Lord, 1980). Because trait levels are inherently unobservable, item response theory falls into the general class of latent trait models (see *Latent Trait Measurement Models*).

In contrast to classical test theory, item response theory makes strong assumptions about a person's behavior when responding to items. Many advantages accrue from these strong assumptions, for example: (a) it is possible to characterize or describe an item, independently of any sample of people who might respond to the item; (b) it is possible to characterize a person independently of any sample of items administered to the person; and (c) it is possible to predict properties of a test in advance of test administration.

Item response theory has some disadvantages. For some models it is currently not possible to check completely the accuracy with which the assumptions are met by the data. For data that appear to meet the assumptions, however, it is reassuring that predictions made from item response theory can often be independently verified. Applications of item response theory are generally more expensive than similar applications of classical test theory, and many applications of item response theory require the use of a computer.

1 Basic concepts of item response theory

1.1 *Assumptions*

Most item response theory models assume that only a single latent trait underlies performance on an item. This is often a reasonable assumption: most tests are constructed to measure a single trait – for example, verbal ability. Models that incorporate more than one latent trait are currently beyond the state of the art.

Item response theory assumes that it is possible to describe mathematically the relationship between a person's trait level and performance on an item. This mathematical description is called an "item response function", an "item characteristic curve", or a "trace line".

1.2 Item response functions

For dichotomously scored items (items that are scored right or wrong), the item response function (IRF) states mathematically the probability of a correct response for a given level of trait. This conditional probability is a function of the item characteristics or parameters. Usually, the mathematical function chosen to represent this conditional probability is from the logistic ogive family or the normal ogive family of functions. There is little difference between the two. More practical work has been done using the logistic family, because of its mathematical simplicity.

If u_i stands for a response to item i (0 for incorrect and 1 for correct) and θ stands for the trait being measured, then the logistic item response function is:

$$P(u_i = 1 \mid \theta) = c_i + (1 - c_i)/(1 + e^{-1.702 a_i (\theta - b_i)}) \tag{1}$$

The normal ogive item response function is:

$$P(u_i = 1 \mid \theta) = c_i + (1 - c_i)\Phi[a_i(\theta - b_i)] \tag{2}$$

where $\Phi[\ ]$ is the normal cumulative distribution function. In these equations a_i, b_i and c_i are parameters that describe characteristics of item i. The pseudoguessing parameter c_i is the probability that an examinee with very low θ will respond correctly to the item. The item discrimination parameter a_i is related to the steepness of the curve at the point of inflection. The item difficulty b_i is the θ-level at the point of inflection.

Not all items require three parameters to characterize them adequately. Some work has been done with two-parameter models ($c_i = 0$). A great deal of work has been done with one-parameter models in which the items vary only in difficulty ($a_i = $ constant, $c_i = 0$). If this latter model is logistic, it is called the Rasch model (see *Rasch Measurement Theory; Rasch Measurement Models*). Note that the three-parameter model in Equation (1) and Equation (2) subsumes models with fewer parameters.

Item response functions from different (logistic) models are displayed in Figure 1. Two three-parameter IRFs are displayed in the bottom panel. It should be noted that both IRFs have nonzero pseudoguessing parameters and the discriminations and difficulties differ. The middle panel displays IRFs from the two-parameters family. Note that both IRFs have zero pseudoguessing parameters and the discriminations and difficulties differ. The top panel displays IRFs from the one-parameter family. They are parallel to each other and differ only by a shift in difficulty or location.

Item response models have also been developed for items with more complex scoring procedures. Consider a multiple-choice item for which it is informative to know which incorrect option was selected by a person. Item response theory applicable to this type of scoring has been developed by Samejima (1969), Bock (1972) and Masters (1982).

1.3   Information functions

Information functions (Birnbaum, 1968) are used to describe the measurement effectiveness of a test or an item at each level of the trait being measured. In contrast, classical test theory usually provides only one measure of effectiveness, which is applied to all people regardless of their θ.

The test information function for a particular scoring method has two equivalent definitions, both of which are useful. By the first definition, the information function for test score y, $I(\theta, y)$, is inversely proportional to the square of the length of the asymptotic confidence interval for estimating trait θ from score y. A high level of information at a

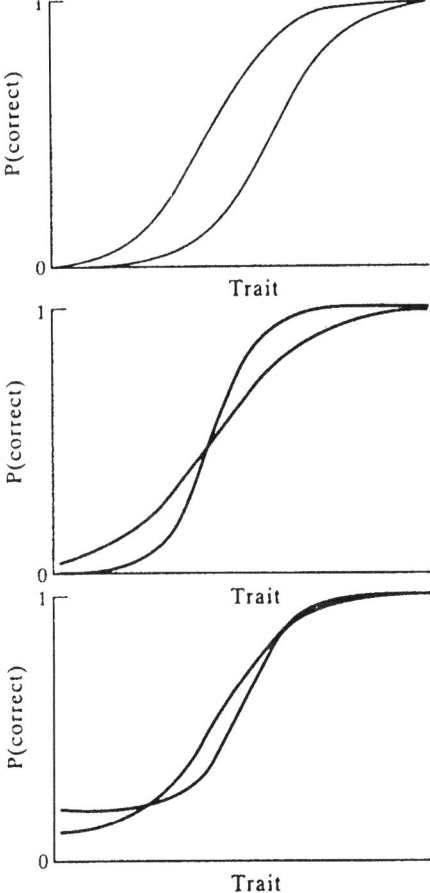

Figure 1   Two one-parameter IRFs (top), two two-parameter IRFs (middle), and two three-parameter IRFs (bottom)

particular θ means that this θ can be more precisely estimated from score y than a θ for which the level of information is relatively low.

The second definition states that the information function for test score y is the square of the ratio of (the slope of the regression of y on θ) to (the standard error of measurement of y for fixed θ). In this context, $I(\theta, y)$ can be viewed as a signal-to-noise ratio. The signal is the change in mean y due to a change in θ. The noise is measured by the standard error of measurement of y for fixed θ.

An item information function $I(\theta, u_i)$ is defined as $I(\theta, u_i) = [P_i'(\theta)]^2/[P_i(\theta)Q_i(\theta)]$, where $P_i(\theta) \equiv P(u_i = 1 | \theta)$ is the item response function, $Q_i(\theta) = 1 - P_i(\theta)$; and $P_i'(\theta)$ is the derivative of the item response function with respect to θ. The test information function $I(\theta)$ is defined as the maximum information available from a test, regardless of the scoring method. The test information function is the simple sum of the item information functions: $I(\theta) = \sum_{i=1}^{n} I(\theta, u_i)$, where n is the number of items in the test. For conventional tests $I(\theta)$ is typically a bell-shaped curve. Each item contributes to $I(\theta)$ independently of all other items in the test. In classical item and test analysis, the contribution of each item to test reliability and test validity depends upon what other items are in the test.

Information functions are useful when the metric established for measuring θ is not subject to challenge. However, slight changes in this metric can drastically alter the shape of an information function and hence the conclusions drawn.

1.4 Relative efficiency functions

If it is supposed that there are two tests, x and y, both measuring the same trait θ, then the relative efficiency (RE) function of test y versus test x, RE (y, x), is the ratio of their information functions at corresponding values of θ: $RE(y, x) = I(\theta, y)/I(\theta, x)$ (Birnbaum, 1968). Most practical applications of item response theory will rely on relative efficiency functions since, unlike information functions, relative efficiency is invariant under any monotonic transformation of the metric used to measure θ.

If $RE(y, x) > 1$ for a particular θ, then test y gives more information than test x at that θ. Relative efficiency functions are useful tools for redesigning existing tests and for investigating novel tests, without actually administering them.

2 Choice of models and estimation methods

The first steps in any application of item response theory to practical problems are to choose a mathematical model for the item response function and to obtain estimates of item parameters and perhaps the θs.

2.1 Choosing a model

The choice of the appropriate model depends predominantly on the types of test questions and the scoring of the test questions.

Consider items that will be scored either correct or incorrect. If such an item is multiple-choice, that is, if it can be correctly answered by guessing, then an item response function containing a pseudoguessing parameter is desirable. If such an item is free-response, that is, if an examinee must produce a response rather than select one from a list of possible responses, then it is less likely to be correctly answered by guessing, and a two-parameter model with no pseudoguessing parameter is appropriate. If free-response items vary in difficulty but not in discrimination, then the one-parameter model is most appropriate.

Similar kinds of considerations affect the choice of the appropriate model when test questions are scored by more complex methods.

In practice, the choice of the appropriate model is not independent of the amount of data available for estimating parameters of that model. In general, the larger the number of parameters in a model, the more data (in the form of examinee responses to items) are required to obtain good estimates of model parameters. In some cases, in the absence of adequate data, estimation errors will be reduced by choosing a less appropriate model with fewer parameters (see Lord, 1983).

2.2   *Choosing an estimation method*

The process of obtaining estimates of model parameters is frequently referred to as "calibration". It is difficult for most models – and impossible for some – to obtain such estimates by hand. Many available computer programs, implementing a variety of statistical approaches to the estimation of parameters of many different models, have been developed over the past 20 years. LOGIST (Wingersky, 1983) implements a joint maximum likelihood approach to estimating item parameters and θs for the one-, two-, and three-parameter logistic item response model. BILOG (Mislevy and Bock, 1983) employs a marginal maximum likelihood approach to the estimation of item parameters for the same models, and a Bayesian approach to the estimation of θs. ASCAL (Vale, 1985) employs a joint maximum likelihood approach with prior distributions imposed for the same models.

These programs are quite general in the sense of handling more than one model, therefore they may be less efficient and informative for a single model than computer programs designed to exploit aspects of that single model. BIGSTEPS (Wright and Linacre, 1991) fully exploits features of the one-parameter model, along with extensions to partial credit models based on the one-parameter model.

3   Illustrative applications

Parameter estimates are fallible approximations to true parameter values. To the extent that the approximations are close to the true parameter values, or appropriate methodology has been employed in order to take account of the uncertainty in the estimates (see Mislevy *et al.*, 1993 and Tsutakawa and Soltys, 1998), item response theory provides for many powerful applications to measurement problems.

3.1   *Test construction*

Tests with prespecified measurement properties can be constructed from a pool of calibrated items. The first step is to specify a target information function for the new test. The shape of this target indicates the θ levels at which the test should provide the most precise measurement. Second, select items for the new test that will fill in areas under the target that might be difficult to fill, for example, areas where relatively few items are available. Third, compute the test information function for this part test. Fourth, add items that contribute information in areas that are far from meeting the target. Continue to choose items, always comparing the information function of the part test to the target, until a satisfactory approximation to the target has been reached.

   Much work has been done to facilitate the solution of IRT-based test design problems using computerized item banks and more complex procedures for selecting test items. Methods of eliciting target information functions and incorporating content and other nonstatistical constraints on item selection have been developed which cast the test design problem as a mathematical programming problem (van der Linden and Boekkooi-Timminga, 1989).

3.2   *Redesigning an existing test*

Relative efficiency functions provide a convenient way of investigating various design changes in a test and comparing them with the original test. Figure 2 illustrates this.

   The curves in Figure 2 are relative efficiency functions for three different tests designed from an original 60-item test. Curve 1 is a test containing only the 30 harder items. This half-length test is less efficient for all test scores. The loss of efficiency is small, however, for high scores. Curve 2 is a test containing only the 30 easier items. This half-length test is less efficient at high scores but is actually more efficient than the full-length test at low scores. This is so because guessing on hard items by low-scoring people destroys

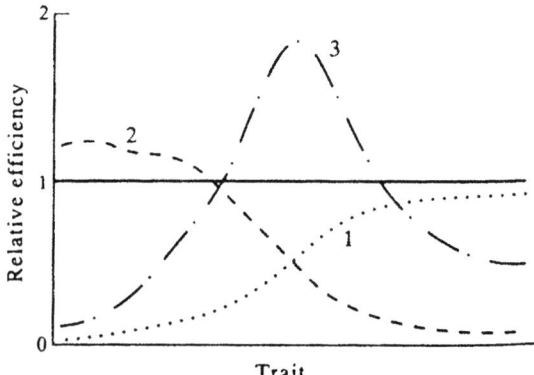

Figure 2   Relative efficiency functions of three modified tests

information. Better measurement is obtained for these people by discarding the harder items. Curve 3 is a 60-item test with all b_i changed to a middle value. This "peaked" test measures very much better for the middle range of scores.

3.3 Equating

Equating is a measurement topic of interest to test publishers who produce many different forms of a test but wish to report scores on a single scale. The process of finding corresponding scores on different forms of a test is called "equating". In general, observed scores on two different forms of a test cannot be equated except under conditions that make the equating unnecessary (Lord, 1980). However, true scores (see *Equating of Tests*) can be equated under a wide variety of conditions, and item response theory facilitates true-score equating.

Suppose there are two tests, x and y, both measuring the same trait. The items in both tests are calibrated on the same scale. A person's (number-right) true score T_x, on test x is a transformation of the person's θ: $T_x = \sum_{i=1}^{n} P_i(\theta)$, where $P_i(\theta)$ is the item response function for item i evaluated at θ, and n is the number of items in text x. Similarly, the true score on test y is T_x, on test x is a transformation of the person's θ: $T_y = \sum_{i=1}^{m} P_i(\theta)$, where the sum is taken over the m items in test y. These two expressions imply that for any particular θ, a true score on test x and a corresponding true score on test y can be computed. These two true scores are equated because they represent the same trait level on two different measuring scales.

In practice, estimated item parameters are substituted into the two expressions, and equated pairs of true scores are computed for arbitrary values of θ. Even though only observed scores are known, the process continues as if the true-score equating holds for observed scores as well.

3.4 Item bias

Items in a test that measures a single trait should measure the same trait in all subgroups of the population to which the test is administered. Items that fail to do so are biased for or against a particular subgroup. Since item response functions in theory do not depend upon the group used for item calibration, item response theory provides a natural method for detecting item bias.

Suppose a test is administered to two different groups and the item response functions are estimated separately for each group. If, for a particular item, the item response function for one group is uniformly higher than for the other group, then a person in the first group has a higher probability of a correct response for the same θ. This item is clearly biased in favor of the first group. Typical instances of item bias are not this clear. Usually item response functions will cross, rather than lie all above or all below each other. This means that the item is differentially biased at different θ levels.

3.5 Mastery testing

Mastery tests are designed to determine if a person has reached a specified level of achievement, in which case the person is a "master". Item response theory can be used to construct optimal mastery tests.

If a pool of calibrated items is supposed, for which subject-matter experts have defined the mastery level, the test construction specialist then selects items from the pool that have the highest item information functions at that level. The test constructed in this way will measure most precisely at the mastery level, thus minimizing errors in classifying people. In addition, item response theory can aid in the determination of the optimal item difficulty, item weights, and cutting score, as well as in the necessary test length. Various test designs may be compared in relative efficiency.

3.6 Tailored testing

Conventional tests are usually designed to measure best near the middle of the θ range for some group. A tailored test is one in which every person is administered items that measure that person's θ best. The general testing algorithm is as follows: (a) obtain an estimate of a person's θ; (b) from a pool of calibrated items, select an item that measures best at that θ; (c) administer and score the item and revise the estimated θ; and (d) if the estimate is precise enough, stop; otherwise, return to step (b). Many tailored testing designs are possible, most of which require a computer for item administration.

Information functions for tailored tests are generally higher than for conventional tests over a broad range of θ. Much work has been done in this field, investigating and implementing various designs (Dorans *et al.*, 1990).

Item response theory is essential for many aspects of tailored testing. For example, conventional scoring does not apply, since every person may take a different test. In contrast, item response theory provides estimates of θ that are independent of the particular items administered. With item response theory, many different designs can be examined and evaluated using relative efficiency functions (see also: *Adaptive Testing; Latent Trait Measurement Models*).

References

Birnbaum, A. 1968. Some latent trait models and their use in inferring an examinee's ability. In: Lord, F. M. and Novick, M. R. (eds.) 1968. *Statistical Theories of Mental Test Scores*. Addison-Wesley, Reading, Massachusetts.

Bock, R. D. 1972. Estimating item parameters and latent ability when responses are scored in two or more nominal categories. *Psychometri.* 37(1), 29–51.

Dorans, N. J. *et al.* 1990. *Computerized Adaptive Testing: A Primer*. Erlbaum, Hillsdale, New Jersey.

Lord, F. M. 1980. *Applications of Item Response Theory to Practical Testing Problems*. Erlbaum, Hillsdale, New Jersey.

Lord, F. M. 1983. Small N justifies Rasch methods. In: Weiss, D. (ed.) 1983. *New Horizons in Testing*. Academic Press, New York.

Masters, G. N. 1982. A Rasch model for partial credit scoring. *Psychometr.* 47(2), 149–74.

Mislevy, R. J. and Bock, R. D. 1983. BILOG: Item analysis and test scoring with binary logistic models (computer program). Scientific Software Inc., Mooresville, Indiana.

Mislevy, R. J., Sheehan, K. M. and Wingersky, M. S. 1993. How to equate tests with little or no data. *J. Educ. Meas.* **30**(1), 57–78.

Samejima, F. 1969. Estimation of ability using a response pattern of graded scores. *Psychometr. Monogr.* Vol. 34, No. 17, Part 2. University of Chicago Press, Chicago, Illinois.

Tsutakawa, R. K. and Soltys, M. J. 1988. Approximation for Bayesian ability estimation. *J. Educ. Stat.* **13**(2), 117–30.

Vale, C. D. 1985. ASCAL: Item parameter estimation program (computer program). Assessment Systems Inc., St Paul, Minnesota.

van der Linden, W. J. and Boekkooi-Timminga, E. 1989. A maxi-min model for test design with practical constraints. *Psychometri.* **54**(2), 237–48.

Wingersky, M. S. 1983. LOGIST: A program for computing maximum likelihood procedures for logistic test models. In: Hambleton, R. K. (ed.) 1983. *Applications of Item Response Theory.* Educational Research Institute of British Columbia, Vancouver.

Wright, B. and Linacre, M. 1991. *A User's Guide to BIGSTEPS.* MESA Press, Chicago, Illinois.

Further reading

Hambleton, R. K. and Swaminathan, H. 1984. *Item Response Theory: Principles and Applications.* Kluwer Nijhoff, Boston, Massachusetts.

Hulin, C. L., Drasgow, F. and Parsons, C. K. 1983. *Item Response Theory: Application to Psychological Measurement.* Dow Jones-Irwin, Homewood, Illinois.

5   Rasch Measurement Theory

P. Allerup

This entry is concerned with the use of the Rasch model as a diagnostic tool in the development of tests in order to identify items and persons that do not conform to the measurement model. It emphasizes the unique properties of the dichotomous Rasch model, and the problems encountered when items do not discriminate at the same level between individuals. This entry is of a technical nature and provides a mathematical and statistical treatment of the issues. This approach has been adopted because although the Rasch models are used increasingly in many parts of the world, texts that raise and address these issues are not readily available. Other entries in this volume consider the applications of the Rasch model in a wide range of measurement problems (see *Adaptive Testing; Essays: Equating of Marks; Item Banking; Partial Credit Model; Rating Scale Analysis*).

1   Introduction to the Rasch model

When the name "Rasch model" appears in connection with statistical analyses, it will often mean the so-called "dichotomous Rasch model" (Rasch, 1960) or the "one-parameter logistic model", namely, the basic statistical model, used when analyzing a set of observations, $a_{vi} = 0$ or $a_{vi} = 1$, obtained from a study, where n students ($v = 1, \ldots, n$) respond (correctly or incorrectly) to k ($i = 1, \ldots, k$) test items. When Georg Rasch (1901–80) was still alive, he strongly opposed the use of "Rasch" as a label for the model – or rather models, since, in fact, the dichotomous model is only one model among a group of statistical models. Rasch himself referred to "models for measurement" or "measurement models".

Among the more prominent members of this group is the "general M-category model" (Rasch, 1968), in which the responses a_{vi} take M different values (e.g., M = 3: "agree"/"do not agree"/"do not know"). The general M-category model is, like the basic Rasch model, a statistical model for analyzing categorical information a_{vi} but the fact that the basic model summarizes the number of correct responses tends to suppress this aspect of the model. Although the multiplicative Poisson model was used as an illustrative example before 1960 by Rao (1952), it was Rasch (1960) who extensively used the multiplicative structure of the expected values $\mathrm{E}(a_{vi}) = \lambda_{vi} = \varepsilon_i \xi_v$ (λ is the expected value of a Poisson variable, splitting multiplicatively into ε and ξ characterizing row and column effects) in his working with the concept of specific objectivity (Rasch, 1966a; 1966b; 1966c) and exploited the statistical properties of this class of models in practical statistical analysis. The general M-category model attained a position as a thoroughly investigated model when Andersen (1973) solved the problems of conditional estimation routines for the parameters and gave proofs for the statistical distributions in large samples of the statistics in a conditional framework. Later, it seems that a particular version of the general model, the so-called "one-dimensional general model", in which the item and individual parameters (originally being arrays containing as many parameters as the number of response categories but now reduced to a set of one-dimensional category and item parameters) (Andrich, 1978; Allerup, 1985, 1986, 1987) has been used at the cost of the general model.

The Rasch model with dichotomous responses $a_{vi} = 0, 1$ quickly became the model attracting most attention, and very soon Fischer (1974) published a comprehensive treatment of the mathematics of this model and the classical statistical methods. However, it seems that some of the powerful properties of this model have been disregarded in the way the model is used in practical data analysis. This occurs in two quite different situations that employ the model. The first is when the model is tested as a theoretical description of data (i.e., the model is tested for fit) but the model is rejected; in this case substantial knowledge is still available from the very way in which the model was rejected. The other situation is where the model is accepted as an adequate description but not all consequences of this acceptance are taken properly into consideration.

The aim of this entry is therefore twofold: to illustrate through examples that: (a) rejection of the Rasch model can in itself represent valuable "end point" information, ready for important conclusive statements; and (b) that when data are further analyzed, acceptance of the Rasch model implies the existence of a set of restrictions on the statistical hypotheses and analyses.

2 Uniqueness of the dichotomous Rasch model

In the examples below it is advocated that fitting data by the Rasch model is, in fact, a one-to-one correspondence between accepting a certain mathematical structure in the data and having available a definite set of practical tools – or interpretations – for the further analysis of data. A standard test of fit of a statistical model is always conducted by means of a (limited) set of test statistics derived under the model hypothesis; and considering the conventional problems of Type-I and II errors the analyst is led to accept or reject the model. Since test statistics are derived mathematical consequences under the model hypothesis, they, by nature, can be "strong" consequences; that is, unambiguously derived consequences. Alternatively they can be "weak" in the sense that this particular test statistic is but one out of several other, important consequences of the model hypothesis.

Table 1   Responses $a_{vi} = 0, 1$ from n individuals to k items with individual parameters ξ_v $v = 1, \ldots, n$ and item parameters ε_i $i = 1, \ldots, k$

		Item No. 1 . . . . i . . . . k			Scores (r_v)
Individuals	1	1 0 1		1	$a_{1.}$
	2	1 1 0	.	0	$a_{2.}$
	.		.		.
	v	. . . . a_{vi}			$a_{v.}$
	.		.		.
	.		.		.
	n		.		.
Item totals (s_i)		$a_{.1}$ . . . . . . $a_{.i}$ . . . . . . $a_{.k}$			$a_{..}$

This is the problem of specificity of the statistics used for testing the statistical model; that is, how much can be inferred about the model from the test statistics? In order to make the position of the Rasch model clear it is necessary to reproduce briefly the proof of specificity of the test statistics, using the original terminology, since Rasch gave this proof only as a note in 1971.

Assume that the set of responses $a_{vi} = 1$ ("correct") or $a_{vi} = 0$ ("noncorrect") are organized in the well-known scheme (see Table 1) with $\xi_1, \ldots, \xi_n$ indicating the individual (latent) parameters and $\varepsilon_1, \ldots, \varepsilon_k$ indicating the item parameters.

The Rasch model, then, assigns the following probability in Equation (1) for obtaining the response a_{vi}

$$P(a_{vi}) = \frac{(\varepsilon_i \xi_v)^{a_{vi}}}{1 + \varepsilon_i \xi_v} \tag{1}$$

$i = 1, \ldots, k; v = 1, \ldots, n.$

From Equation (1) the probability of any matrix $((a_{vi}))$, conditional upon the marginals, is calculated

$$\sum_{i=1}^{k} a_{vi} = r_v \sum_{v=1}^{n} a_{vi} = s_i. \tag{2}$$

$r_v \, v = 1, \ldots, n$ are the individual scores, $s_i \, i = 1, \ldots, k$ are item totals, and show that this probability is independent of all the parameters ξ_v, ε_i.

In fact,

$$\begin{bmatrix} (r_v) \\ (s_i) \end{bmatrix}$$

denoting the number of zero-one matrices with the marginals in Equation (2), it is possible to obtain, readily, the conditional probability

$$p(((a_{vi}))|(r_v), (s_i)) = \frac{1}{\left[\begin{array}{c}(r_v)\\(s_i)\end{array}\right]}$$

(3)

Since Equation (3) is dependent on the marginals (r_v) and (s_i) only, it shows, furthermore, that all such matrices are equally probable.

That Equation (3) proves true, then, is a necessary condition for the model to be true; thus, by means of the statistics (r_v) and (s_i) it offers an opportunity for test of fit of the model for an actual set of data.

However, in the following it can be proved that Equation (3) is also a sufficient condition for the model in Equation (1), provided the a_{vi}s are stochastically independent.

No assumptions are required to write

$$p(a_{vi} = 1) = \frac{\lambda_{vi}}{1 + \lambda_{vi}}$$

$$p(a_{vi} = 0) = \frac{1}{1 + \lambda_{vi}} \quad \lambda_{vi} > 0$$

(4)

and it can be concluded from the independence of the a_{vi}s that the probability generating function with the associated variables $((X_{vi})) = x$, for $a = ((a_{vi}))$ is

$$\prod \left[\begin{array}{c}a\\x\end{array}\middle|\lambda\right] = \prod \left[\left(\begin{array}{c}a_{vi}\\x_{vi}\end{array}\right)\middle|((\lambda_{vi}))\right] = \prod_{(v)}\prod_{(i)} \left(\frac{1 + \lambda_{vi}X_{vi}}{1 + \lambda_{vi}}\right)$$

(5)

If the associated variables $x_{vi} = x'_{vi}$, y_v, z_i are factorized, and using the notation

$$y_r = y_{r_1}, \ldots, y^{r_n} \qquad x^a = \prod_{(v)}\prod_{(i)} x_{vi}^{a_{vi}}$$

$$z^s = z^{s_1}, \ldots, z^{s_k} \qquad \lambda^a = \prod_{(v)}\prod_{(i)} \lambda_{vi}^{a_{vi}}$$

(6)

$$r = (r_1, \ldots, r_n) \qquad s = (s_1, \ldots, s_k)$$

the generating function in Equation (5) can be written

P. Allerup

$$\pi \begin{bmatrix} a \\ x \end{bmatrix} \lambda \end{bmatrix} = \sum_{(r,\,s)} \prod \begin{bmatrix} a \\ x' \end{bmatrix} r,\,s,\,\lambda \end{bmatrix} \cdot p(r,\,s\,|\,\lambda)\,y^r z^s \tag{7}$$

Now, each factor in the generating function in Equation (5) may be transformed into one, with the parameter $\lambda = ((1))$. In fact,

$$\prod \begin{bmatrix} a_{vi} \\ x_{vi} \end{bmatrix} \lambda_{vi} \end{bmatrix} = \frac{1 + \lambda_{vi} x_{vi},\,1}{1 + 1} \cdot \frac{2}{1 + \lambda_{vi}} = \prod \begin{bmatrix} a_{vi} \\ \lambda_{vi} x_{vi} \end{bmatrix} 1 \end{bmatrix} \cdot \frac{2}{1 + \lambda_{vi}} \tag{8}$$

Thus, using the notation $u \circ v = ((u_{vi}\,v_{vi}))$, which is valid for any two response matrices of the same order $(n,\,k)$, and

$$((x'_{vi}\,y_v\,z_i)) = x' \circ (y*z)$$
$$((\lambda_{vi}\,x'_{vi}\,y_v\,z_i)) = \lambda \circ x' \circ (y*z) \tag{9}$$

then, Equation (8) applied to Equation (5) leads to the identity

$$\prod \begin{bmatrix} a \\ x \circ (y*z) \end{bmatrix} \lambda \end{bmatrix} = \prod_{(v)} \prod_{(i)} \cdot \frac{2}{1 + \lambda_{vi}} \cdot \prod \begin{bmatrix} a \\ \lambda \circ x' \circ (y*z) \end{bmatrix} ((1)) \end{bmatrix} \tag{10}$$

Using the conditional approach in Equation (7), with powers of y and z for both sides of Equation (10) and identifying corresponding terms, the following general relation is obtained

$$\prod \begin{bmatrix} a \\ x' \end{bmatrix} r,\,s,\,\lambda \end{bmatrix} p(r,\,s\,|\,\lambda) = \prod_{(v)} \prod_{(i)} \cdot \frac{2}{1 + \lambda_{vi}} \cdot \prod \begin{bmatrix} a \\ \lambda \circ x \end{bmatrix} r,\,s,\,((1)) \end{bmatrix} \cdot p(r,\,s\,|\,((1))) \tag{11}$$

For the particular pairs of observed marginals $(r,\,s)$ Equation (11) can, through the development of Equation (5) to Equation (7), be written as

$$\frac{\varphi(\lambda \circ x'\,|\,r,\,s)}{\varphi(x'\,|\,r,\,s)} = \prod_{(v)} \prod_{(i)} \frac{1 + \lambda_{vi}}{2} \cdot \frac{p(r,\,s\,|\,\lambda)}{p(r,\,s\,|\,((1)))} \tag{12}$$

using that

$$\varphi(x'\,|\,r,\,s) = \sum x'^a \tag{13}$$

the summation being extended over all a_{vi}s with fixed $(r,\,s)$.

Since the right-hand term of Equation (12) is independent of x', the left-hand term, for any x', is identical with what is obtained by putting $x' = ((1))$. Consequently, the function φ satisfies the following functional equation

$$\varphi(\lambda \circ x' \mid r, s) = \frac{\varphi(\lambda \mid r, s)}{\begin{bmatrix} r \\ s \end{bmatrix}} \cdot \varphi(x' \mid r, s) \tag{14}$$

If Equation (13) is applied on both sides of Equation (14), using that $(\lambda \circ x')^a = \lambda^a \circ x'^a$, the following can be obtained immediately

$$\lambda^a = \frac{\varphi(\lambda \mid r, s)}{\begin{bmatrix} r \\ s \end{bmatrix}} \tag{15}$$

This is valid for any response matrix $((a_{vi}))$ with the marginals (r, s). It follows, that for any two response matrices a and a', yielding the given marginals (r, s):

$$\lambda^a = \lambda^{a'} \Leftrightarrow \sum_{(v)} \sum_{(i)} a_{vi} \kappa_{vi1} = \sum_{(v)} \sum_{(i)} a'_{vi} \kappa_{vi1} \quad \kappa_{vi} = \log \lambda_{vi} \tag{16}$$

For any two admissible matrices a and a', being equal for all (v, i), except for a so-called "switch" in an intersection between two columns (i, j) and two rows (u, v):

a	a'
1 0	0 1
0 1	1 0

Equation (16) requires that

$$\kappa_{ui} + \kappa_{vj} = k_{uj} + \kappa_{vi} \tag{17}$$

A final averaging in Equation (17) over the indices v and j leads to

$$\kappa_{ui} = \kappa_{u.} + \kappa_{.i} - \kappa_{..} \tag{18}$$

which means that $\lambda_{ui} = \log \kappa_{vi}$ splits into a u-factor (row) and an i-factor (column). Inserting this factorization of λ into Equation (4), completes the proof that (3) is also a sufficient condition (see *Sufficient Statistics in Educational Measurement*).

2.1   Consequences of the uniqueness

The proof states that the Rasch model in Equation (1) is true if – and only if – the conditional distribution of the responses $((a_{vi}))$ given the marginals (r_v) and (s_i) is independent of the parameters $\varepsilon_1, \ldots, \varepsilon_k$ and $\xi_1, \ldots, \xi_n$ – briefly denoted as joint sufficiency of the marginals (r_v) and (s_i).

Before the conclusions from the proof of uniqueness of the Rasch model are fully drawn, it is necessary to consider another aspect of the model's uniqueness in terms of specific objective comparisons. In fact, Rasch (1966a, 1966b, 1966c) along with the development of statistical properties of the model, like the theorem of specificity shown above, was deeply attached to the problem of defining the general conditions under which individuals in Table 1 can be compared – or measured. The question to be solved was essentially the following: What are the formal requirements for being able to compare any two individuals v and u – through the parameters ξ_v and ξ_u – using any subset of the items – namely the item parameters – and end up with (stochastically) the same result? This formulation is, of course, an abbreviated form of Rasch's original request for objectiveness when comparing individuals. In one of his last papers Rasch (1977) summarized his work with specific objective comparisons and stressed that general objectivity seems to be an ambiguous concept, and hence objectivity needs a frame of reference – like the one given by Table 1 – in order to be defined distinctly. Therefore, "specified" or "specific objective comparisons" among individuals (and among items) are the terms used by Rasch in order to restrict, or specify, the comparisons to be undertaken in such frameworks as Table 1.

It turns out (Rasch, 1968, 1977) that the request for specific objective comparisons is, in fact, in one-to-one correspondence with the Rasch model in Equation (1) – if response matrices $((a_{vi}))$ with $a_{vi} = 0, 1$ like the Table 1 data are considered.

Now, combining the conclusions from the proof of uniqueness and the properties of objectivity it is seen that:

{The Rasch model (1) is true}
if, and only if
{The marginals (r_v) and (s_i) are jointly sufficient statistics}
if, and only if
{Individuals (items) can be compared (specific) objectively}

The mutual characterization of (a) the statistical model, (b) the sufficiency of marginals, and (c) specific objective comparisons have a series of implications for the practical work. In fact, (b) enables the analyst – in principle by applying the test statistic in Equation (3) above – to carry out a test of fit of the model that is independent of all parameters; and to do it specifically. If the model is rejected, the individual total scores (r_v) are no longer sufficient statistics, and information about the ξ's is not exhausted by the r_v's. In other words, information is needed about which items have been answered correctly instead of knowing only the number of correct responses across all items. The way the test statistic rejects the model indicates the "direction" of misfit and, taking into account the specificity of the test statistic, "exactly" what is wrong is now known. Usually, the items are so-called nonhomogeneous, and proper use of the test statistics may point to, for example, the existence of two homogeneous subscales of the items rather than one. Rejection of the model is also equivalent to the fact that individuals – in terms of the ξs – cannot be compared, irrespective of which items are used. That is, of course, not the same as claiming that individuals cannot be compared at all, but in this case, any statement concerning two individuals will be nonobjective in the sense that the result of the comparison will depend on something less "general" than what is covered by the whole range of items (e.g., the concept of "difficulty" valid for the items in many test forms).

3   Detecting gender differences

When test forms are analyzed, frequently a sample of boys and girls have answered a series of tasks, or items, which can be classified as correct ($a_{vi} = 1$) or incorrect ($a_{vi} = 0$). The complete set of observations is then gathered in a scheme like Table 1. In the Reading Literacy Study conducted by the International Association for the Study of Educational Achievements (IEA) (Elley, 1992) the items were *a priori* collected in three subgroups, dealing with the reading of three different text types, but still the items could be more or less "difficult" within each reading domain. The caption applied to all items (viz., the ε s) is, therefore, a concept of difficulty and, consequently, "reading comprehension" or "reading ability" is the derived concept assigned to the ξs of the individuals.

Part of the international calibration exercise, where one unique set of items acts as an international ruler, or Reading Scale, for comparing all students, is a test of fit of the Rasch model in Equation (1), in which special attention is given to test statistics revealing possible inhomogeneities between girls and boys. The issues are not that of investigating differences in the level of achievement – which can be studied after acceptance of the model by comparing the ξs for the girls and the ξs for the boys; rather, the question is raised whether the items are equally difficult for the girls and for the boys – irrespective of the general level of ability for the two sexes. In other words, are the item parameters $\varepsilon_1, \ldots, \varepsilon_k$ consistent across all students? Only in this case, of course, can the levels of achievement be compared properly without giving free score points to either of the gender groups because of inhomogeneity. Classical statistical analyses mix up these two aspects of measuring differences, since it is outside their mathematical formalization to distinguish between ability and difficulty in the way the Rasch model does. In order to demonstrate how test of fit of the Rasch model also brings about conclusive statements concerning gender differences, it is necessary, first of all, to take advantage of the sufficient reduction of the basic observations $((a_{vi}))$ conducted by the r_vs.

This is done in Table 2, where the individuals are classified according to their score values r_v, $v = 1, \ldots, n$. If the conditional approach is employed with the test statistic in Equation (3), the conditional probability can be calculated to achieve a correct response to Item No. i both for the boys and for the girls. Imagine that Table 2 summarizes data for the boys and the girls separately – $a_{\cdot i}(r)$s and $b_{\cdot i}(r)$s, referring to the boys and the girls

Table 2   The number of $a_{.i}(r)$ of correct responses to Item No. i given by n_r individuals in score group $= r$ [a]

Score r	Item No. $1 \ldots i \ldots k$	Number of individuals in score group n_r
0	$0 \ldots 0 \ldots 0$	n_0
.	.	.
r	$\ldots a_{.i(r)} \ldots$	n_r
.	.	.
$k-1$		n_{k-1}
k	$n_k \ldots n_k \ldots n_k$	n_k
Total	$s_1 \ldots . s_i \ldots s_k$	$n = \Sigma n_r$

[a] The total number of correct responses to Item No. i is $s_i i = 1, \ldots, k$

respectively; and, further, n_r and m_r indicate the number of boys and girls in score group: r correct responses.

From the Rasch model in Equation (1) the distributions of $a_{.i}(r)$ and $b_{.i}(r)$ are derived as two binomials:

$$a_{.i}(r) \sim b(n_r, \theta_{ri}) \theta_{ri} = \varepsilon_i^b [\gamma_{r-1}^{(i)} / \gamma_r]_b$$

$$a_{.i}(r) \sim b(m_r, \eta_{ri}) \eta_{ri} = \varepsilon_i^g [\gamma_{r-1}^{(i)} / \gamma_r]_g$$

ε_i^b and ε_i^g are the difficulties for Item No. i for boys and girls respectively, and γ_r is the elementary symmetric function of $\varepsilon_1, \ldots, \varepsilon_k$ of order r (Rasch, 1960; Fischer, 1974) – $\gamma_{r-1}^{(i)}$ calculated without ε_i and suffixes b and g set to indicate that the γs are calculated for the boys' and for the girls' εs respectively. Notice that the individual parameters ξs are eliminated! From Equation (19) the extended Hypergeometric probability in Equation (20) of $a_{.i}(r) = x$, conditional on $a_{.i}(r) + b_{.i}(r) = c_{.i}(r)$ is derived for each score level r:

$$p(a_{.i}(r) = x \mid c_{.i}(r) = c, n_r, m_r) = \frac{\binom{n_r}{x} \binom{m_r}{c-x} \kappa_{ri}^x}{\sum_{j=0}^{c} \binom{n_r}{j} \binom{m_r}{c-j} \kappa_{ri}^j}$$

(20)

$$\kappa_{ri} = \frac{\theta_{ri}}{1 - \theta_{ri}} \Big/ \frac{\eta_{ri}}{1 - \eta_{ri}} = \varepsilon_i^b / \varepsilon_i^g$$

Under the hypothesis that Item No. i is equally difficult for boys and girls, $H : \varepsilon_i^b = \varepsilon_i^g$, it is observed that $\kappa_{ri} = 1$ *for each score value $r = 1, \ldots, k - 1$.*

Evaluation of the test statistics $a_{\cdot i}(r)$ in Equation (20) for $r = 1, \ldots, k - 1$ gives information on possible deviations from the hypothesis H and — using approximate $\chi^2(1) = u_r^2$ for the standardized $u_r = [a_{\cdot i}(r) - E(a_{\cdot i}(r)|H)] / \sigma(a_{\cdot i}(r))$ at each level of the score $= r - a$ summary test, $\chi^2(df) = \Sigma\, u_r^2\ df = k - 1$ across all score groups is obtained as an overall test for no gender differences – or sex bias – on item No. i.

As an example, Table 3 shows data from Items A and B. In Item A the hypothesis H is rejected, $p = 0.02$ and the patterns of $a_{\cdot i}(r)$ and $E(a_{\cdot i}(r))$ clearly demonstrate an interaction between sex and ability, since the sign of the differences $a_{\cdot i}(r) - E(a_{\cdot i}(r)|H)$ shifts for low rs to high rs. For Item B the hypothesis H is also rejected ($p = 0.00$), but here the situation is different, as a general sex bias is observed across all score groups, and in this case the degree of sex bias can be assessed, simply from the estimated value of K_r in Equation (20), or $\log(\kappa_{ri}) = \log(\varepsilon_i^b) - \log(\varepsilon_i^b) \approx 0.14$, if a logarithmic interval scale is used for the difficulties.

4 Analysis of item discrimination

Another area of interest when applying Item Response Theory (IRT) models, is covered by the so-called "two-parameter logistic model" (Lord and Novick, 1968) which is often cited as a generalization of the basic Rasch model in Equation (1). The practical need for formally extending the mathematical structure of the Rasch model appears when the statistical analysis of actual data fails to accept the Rasch model in Equation (1), and various indications as to why the model does not fit lead to the proposal of the two-parameter model is Equation (21):

Table 3 Observed $a_{\cdot i}(r)$ and expected $E(a_{\cdot i}(r))|H$ responses under the hypothesis H of equal item difficulties for boys and girls on two items A and B[a]

r	n_r	m_r	Item A				Item B		
			$a_{\cdot i}(r)$	$E(a_{\cdot i}(r)|H)$	u_r	$a_{\cdot i}(r)$	$E(a_{\cdot i}(r)|H)$	u_r
0	0	0	0	0.0	.	0	0.00	.
1	4	8	0	0.0	.	0	0.00	.
2	16	32	4	4.00	0.00	10	5.00	3.27
3	69	86	22	17.81	1.40	33	27.15	1.93
4	121	159	40	51.86	− 2.88	83	76.06	1.73
5	193	118	91	101.15	− 2.37	176	168.80	2.54
6	89	57	89	89.00	.	89	89.00	.
			$\chi^2(4) = 15.87$			$\chi^2(4) = 23.86$		
			$\chi^2(p = 0.00)$			$\chi^2(p = 0.02)$		

[a] n_r, $m_r =$ the number of boys and girls respectively in score group $= r$
$u_r =$ approximate normal test statistic for each score level

$$P(a_{vi}) = \frac{e^{[\delta_i \theta_i + \sigma_v]a_{vi}}}{1 + e^{\sigma_i(\delta_i + \sigma_v)}} \qquad (a_{vi} = 0, 1)$$

$$\theta_i = \log(\varepsilon_i) \qquad\qquad\qquad\qquad (21)$$

$$\sigma_v = \log(\xi_v)$$

In Equation (21) logarithmic scaling is applied with θs as item parameters and σs as individual parameters. To many users of classical test theory the concept of item discrimination – and therefore a model as in Equation (21) – is connected with evaluation of biserial correlations between responses to an item a_{vi} and the total scores $r_{v.} = \Sigma a_{vi}$ ($v = 1, \ldots, n$). To users of IRT models, the Item Response Functions or Item Characteristic Curves (ICC) with varying slope values interpreted as varying item discriminations, have for a long time (Lord, 1980) been a natural part of the model construction. From this point of view the model in Equation (21) is, indeed, an evident, general framework for analyzing response data.

Obviously, when setting the item discrimination parameters $\delta_i = 1$ $i = 1, \ldots, k$ the Rasch model is obtained, formally, as a special case of the model in Equation (21). But what is really generalized from the Rasch model in Equation (1) to the model in Equation (21) beyond the mathematics? And what is not? First, it is impossible to give independent interpretations of the item and discrimination parameters; this can be seen by considering the mathematical structure of the model in Equation (21) and trying to identify the components δ_i, θ_i and σ_v – which control the probability completely. Since θ_i and σ_v interact additively, a constrained form, using $\theta_i + \sigma_v + \rho$ with $\theta_. = \sigma_. = 0$ (i.e., average = zero), determines these two sets of parameters, but still the combined term $\theta_i + \sigma_v + \rho$ interacts multiplicatively with the δ_i. Hence, by "shrinking" – that is, dividing the term by some constant – the combined scale of difficulty and ability $\theta_i + \sigma_v + \rho$, it is possible to "compensate" by multiplying the δ_is by the same constant in order to maintain the status quo, as regards the final value of the probability $p(a_{vi})$. Second, the discrimination parameters $\delta_1, \ldots, \delta_k$ are always unknown, and since this brings the model in Equation (21) outside the class of exponential distributions, it implies severe estimation problems of the parameters, including the δs. Third, how can individuals be compared under the model in Equation (21)?

In Figure 1 two ICC's with different slopes are displayed, and on the scale of ability (x-axis) two individuals with different levels of ability σ_1, σ_2 judge the difficulties of the two items from their chances of solving the two items correctly. It can be seen directly from the figure that their ranking of difficulties depends on the positions σ_1, σ_2 of the two individuals; the low-achieving σ_2 disagrees with the high-achieving σ_1. In other words, these two items cannot be compared objectively in terms of level of difficulty!

However, if the model in Equation (21) is not used as a formal basis for investigating the data, because of the drawbacks described, the question arises as to how to make inferences about the phenomenon brought about by varying the slopes of the ICC's. The varying slopes are usually observed already in the initial phase of the testing of the model (Allerup, 1994), when the sufficiency of the r_vs is investigated and the empirically observed chances (estimated conditional ICC) of getting a correct response $a_{.i}(r)/n_r \approx \Theta_{ri}$ (compare Equation (19)) is studied as a function of r, $r = 1, \ldots, k-1$.

Since the probability of a correct response to Item No. i, conditional on the individual score $a_{vi} = r$, is given by

$$p(a_{vi} = 1 \mid a_{v.} = r) = \varepsilon_i \frac{\gamma_{r-1}^{(i)}}{\gamma_r} = \frac{e^{\theta_i + \log(\gamma_{r-1}^{(i)}/\gamma_r^{(i)})}}{1 + e^{\theta_i + \log(\gamma_{r-1}^{(i)}/\gamma_r^{(i)})}}$$

(22)

$$\theta_i = \log(\varepsilon_i)$$

it can be identified with the unconditional probability of the Rasch model Equation (1), using logarithmic scaling $\theta_i = \log \varepsilon_i$ $\sigma_v = \log \xi_v$

$$p(a_{vi}) = \frac{e^{(\theta_i + \sigma_v)a_{vi}}}{1 + e^{\theta_i + \sigma_v}} \qquad (a_{vi} = 0, 1)$$

(23)

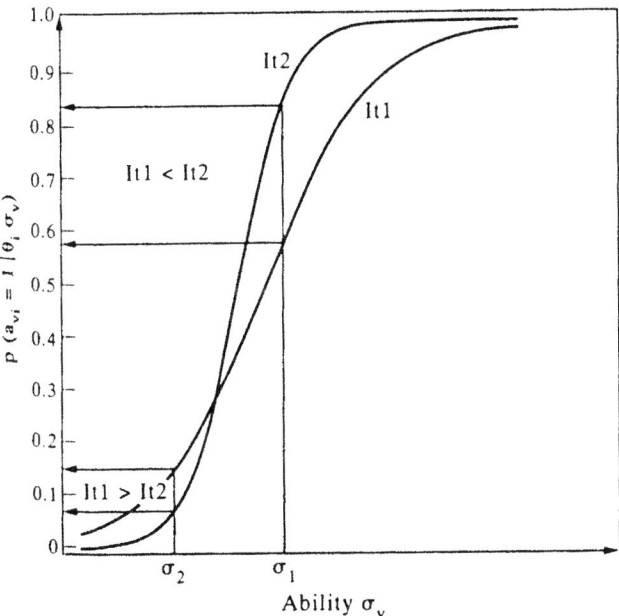

Figure 1 Item characteristic curves (ICC) for items It1 and It2 under the two-parameter model. Item difficulties evaluated ($a > b \Leftrightarrow$ "a easier than b") at two points $\sigma_1 \sigma_2$ on the scale of ability.

and the two-parameter extension of Equation (23) in Equation (21) can be captured directly from Equation (22) by the following model proposal

$$p(a_{vi}|_{v.} = r) = \frac{e^{(\theta_i + \delta_i \lambda_{ri})a_{vi}}}{1 + e^{\theta_i + \delta_i \lambda_{ri}}} \qquad (a_{vi} = 0, 1)$$

(24)

$$\lambda_{ri} = \log(\gamma_{r-1}^{(i)}/\gamma_r^{(i)})$$

The reason for suggesting (a) $\theta_i + \delta_i \lambda_{vi}$ rather than (b) $\delta_i(\theta_i + \lambda_{ri})$ — which would otherwise make the equivalence to the two-parameter model in Equation (21) clearer — is that Equation (24) immediately represents a simple one-dimensional logistic regression (LR) of a_{vi} on the λ's for fixed Item No. i ($r = 1, \ldots, k-1$) with slope $= \delta_i$ and intercept $= \theta_i$. Considering $\lambda_{ri} = \lambda_{ri}(\theta_1, \ldots, \theta_k)$ as known by the estimated θs, the λs being by-products of the conditional maximum likelihood equations for $\theta_1, \ldots, \theta_k$ (Andersen, 1973), tests of fit of LR can be conducted. Besides the usual likelihood ratio test (χ^2) for the shape of the conditional ICC (CICC) modeled by Equation (24), the hypothesis H : $\delta_i = 1$ attracts attention, since $\delta_i = 1$ $i = 1, \ldots, k$ brings the model Equation (24) back to the basic Rasch model in Equation (1). Furthermore, the "locally" LR-estimated intercept values θ_i can be compared with the "global" $\theta_i = \log(\varepsilon_i)$ from the initial estimation of the item difficulties (Allerup, 1994). Notice, that Form (b) can be obtained from Form (a) by the transformation $\theta_i \rightarrow \theta + r_i = \theta_i/\delta_i$ (compare the remarks concerning independent interpretaions of item difficulty and item discrimination in the two-parameter model). It is also worthy of note that the definition of item discrimination δ_i takes place in the model in Equation (24), which — like the basic model in Equation (22) — is conditional on the sufficient statistics $a_{v.} = r$ for the individual parameters $\sigma_1, \ldots, \sigma_n$ under the original Rasch model. This avoids the mathematical confusion caused by the use of $T_v = \Sigma_{(i)} \delta_i a_{vi}$ which, on known δs, is sufficient for the individual parameter σ_v in the two-parameter model in Equation (21). An example with two items showing the results of analyzing the conditional ICC's by means of the logistic regression technique is displayed in Table 4.

It is seen from Table 4 that for Item No. B the test of fit rejects the hypothesized structure in Equation (24) ($p = 0.002 < 0.05$) and, consequently, the ICC for this item not only falls outside what is predicted by the Rasch model, it also fails to conform with simple linear transformations of the basic logistic "S"-shape. For this reason no additional estimation of slope and intercept values takes place. From inspecting the differences between observed and expected frequencies no clear picture emerges of the reason for nonfit; frequently, for the low-scoring individuals, an overweight of observed corrects, namely, the $a_{.i}(r)$s, is a sign of guessing. In Item No. A the situation is different, since test of fit of the item characteristic curve for this item does not reject the structure in Equation (24). The estimated item discrimination $\delta_A \approx 0.83$ and the approximate t-test for the hypothesis $\delta_A = 1$ ($t = -2.75$, $p = 0.02$) indicates that the ICC for item No. A also does not fit the Rasch ICC. However, considering the acceptance of fit to a linear transformation of the basic Rasch ICC, namely, the model in Equation (24), it is desirable to look for other

Table 4 The frequency of correct responses $h_{ri} = a_{\cdot i}(r)/n_r$ within score group $= r$ out of n_r individuals in score group $= r$ [a]

r	n_r	Item no. $i = A$			Item no. $i = B$		
		$a_{\cdot i}(r)/n_r$	$E(h_{ri}\|LR)$	λ_{ri}	$a_{\cdot i}(r)/n_r$	$E(h_{ri}\|LR)$	λ_{ri}
1	2	0.00	0.10	-2.85	0.00	0.07	-2.88
2	5	0.20	0.18	-1.96	0.20	0.25	-2.00
3	12	0.30	0.45	-1.36	0.08	0.19	-1.40
4	6	0.33	0.27	-0.86	0.00	0.13	-0.91
5	33	0.43	0.54	-0.41	0.21	0.31	-0.48
6	46	0.58	0.62	0.01	0.39	0.45	-0.07
7	91	0.67	0.36	0.43	0.32	0.38	0.34
8	183	0.69	0.70	0.86	0.54	0.52	0.77
9	376	0.82	0.78	1.34	0.65	0.61	1.24
10	588	0.86	0.85	1.91	0.75	0.70	1.82
11	915	0.91	0.92	2.76	0.77	0.81	2.67

Test of fit: $\chi^2(9) = 16.15$ ($p = 0.06$) $\qquad\qquad\qquad\qquad \chi^2(9) = 26.50$ ($p = 0.002$)
LR-estimates: $\delta_A \approx 0.83 \qquad \theta_A \approx 0.14$
Approx t: $\quad t = -2.75 \qquad t = 0.10$
$\qquad\qquad (p = 0.02) \qquad (p = 0.92)$

[a] $E(a_{\cdot i}(r)\|n_r/LR)$ the expected frequency of correct responses under logistic regression (LR) and $\lambda_{ri} = \log(\gamma_{r-1}^{(i)}/\gamma_r^{(i)})$ the independent variable in logistic regression; two items, Nos. A and B are studied

items that enjoy similar δ values. In fact, collecting a series of items of this type, it is possible to build up a subscale now fitting the basic Rasch model in Equation (1) – since the common δ_i value, then, would be absorbed into the difficulty parameter θ_i.

If, for various reasons, items satisfying the model in Equation (24) are employed with varying δ_i values – some of which are significantly different from unity – it should then be realized that the estimate of ability $= \sigma_v$ is based upon the statistic $T_v = \Sigma_{(i)} \delta_i a_{vi} (a_{vi} = 0, 1)$ which submits the sum of item discriminations across those items answered correctly. Increasing T_v values – and ability σ_v estimates – are therefore a consequence of either responding correctly to items with high δ's, possibly being easy items with low θ_i's, or responding correctly to an increasing number of items. This again illustrates the confusion in the two-parameter model Equation (21) about the interpretation of the ability σ_v which, in the basic Rasch model, is unambiguously combined with the number of correct answers a_v.

5 Comparing the results of two tests

In Section 3 it was demonstrated how the Rasch model can be used both to detect sex bias on a single item and to assess, quantitatively, the difference by means of the parameters of the model. Here, principally the same issue is addressed, but this time the point of departure lies in the data usually available to users after testing sessions: two sets of individual score values (i.e., the number of correct responses, a_v and b_v, $v = 1, \ldots, n$) –

Table 5 Data from two parallel Tests I and II

Individual No. v	Test I Items 1 $k1$	Test I Score $a_{v.}$	Test I Ability ξ_v	Test II Items 1 $k2$	Test II Score $b_{v.}$	Test II Ability η_v
1	1 0 1 1 0 0	12	1.23	1 1 0 0 1 1	15	1.32
2	0 1 1 0 1 1	13	1.27	1 1 1 0 1 1	18	1.36
.	. . . . . .	.	.	. . . . . .	.	.
v	$((a_{vi}))$			$((b_{vi}))$		
.	. . . . . .	.	.	. . . . . .	.	.
n	1 1 0 1 0 0	9	0.98	1 0 1 1 0 0	8	0.95

[a] Single item responses a_{vi} (Test I) and b_{vi} (Test II) and individual abilities ξ_v, $\eta_v v = 1, \ldots, n$ under the basic Rasch model Equation (1)

one from Test (I) and one from Test (II), displayed in Table 5. It is assumed that the two-test sessions are parallel, so that each individual has match-paired scores from the two tests. Furthermore, it is assumed that the Rasch model proves true for data within each of Tests I and II. This can be done either as a result of performing two separate tests of fit of the Rasch model, conducted at the single-item level, or, in the absence of single-item data, because of the underlying theorem of uniqueness of the model and the sufficiency of the item and individual scores.

In the light of this design a general question is considered: What kind of calculations can be undertaken using the scores $a_{v.}$, $b_{v.}$ in order to reveal the difference betweeen Test I and Test II? One point of view could be that, having accepted the fit of the Rasch model and, consequently, being in a position where all information concerning the individual parameters σ_v, η_v are kept in the scores $a_{v.}$, $b_{v.}$, there is freedom to choose any kind of subsequent calculations based on $a_{v.}$ and $b_{v.}$. In experimental psychology, for instance, Tests I and II (comprising words and nonwords) are used to identify certain reading literacy problems by taking notice of the magnitude of the difference between Tests I and II – the greater the difference, the more distinct is the impression of reading problems; and the more confident is the psychologist about using the difference between Test I and Test II outcomes for diagnostic purpose. The aim of analysis is therefore, in a way, "opposite" to the usual focus, since the statistical analysis is now directed toward identification of those individuals having the greatest deviation from – or rejection of – the hypothesis $H : \sigma_v = \eta_v$.

Numerous suggestions have been published as to how the difference between the two tests can be assessed. The simple algebraic difference (Mitterer, 1982) $a_{v.} - b_{v.}$ between the number of correct responses, a simple quotient $a_{v.}/b_{v.}$ (Bryant and Impey, 1986), a difference between standardized z-scores (i.e., $(a_{v.} - a_{.})/\sigma(a_{v.})$ considering all n individuals) have been proposed (Olson, 1985) and even the so-called d, of signal detection theory (Snowling, 1980), which is the distance (scaled in σ-units) between two mean values from normal distributions by means of cut-off points equal to the percentage of correct responses for Tests I and II. It is, however, easy to show that if these four ways of evaluating the difference between Tests I and II are used, it is possible to end up with incompatible results, leading to a contradictory rank order of the individuals.

There is also the problem of which method to choose. Is one method "better" or more "correct" than the other? The point is that, in fact, none of the methods mentioned is "correct", since none of the methods has properly taken care of the initial assumptions that tacitly lie behind all the methods. These are the use of individual scores as measures of ability and – although not explicit here – the use of item totals as measures of item difficulty. In other words the statistical sufficiency of item and individual scores, according to the theorem of uniqueness, inevitably leads to the Rasch model in Equation (1). What needs to be done, consequently, is to make explicit the use of the Rasch model in Equation (1), apply it to data available from the two tests, and attempt to develop a model-derived statistic, which can be used to assess the individual differences between Tests I and II.

Let a distinction be made between two cases: (a) where all single-item responses $((a_{vi}))$ and $((b_{vi}))$ (and hence by calculation the individual scores $a_{v.}$, $b_{v.}$ $v = 1, \ldots, n$) are known; and (b) where only the individual scores $a_{v.}$, $b_{v.}$ are known.

In both cases (a) and (b) the distribution of individual score values $a_{v.}$ and $b_{v.}$ can be obtained readily from the basic Rasch model in Equation (1)

$$p(a_{v.} = x) = \frac{\xi_v^x \gamma_x(\varepsilon)}{\displaystyle\sum_{j=0}^{k1} \xi_v^j \gamma_j(\varepsilon)}$$

$$p(b_{v.} = y) = \frac{\eta_v^y \gamma_j(\eta)}{\displaystyle\sum_{j=0}^{k2} \eta_v^j \gamma_j(\eta)} \tag{25}$$

$$\gamma_x(\underline{\varepsilon}) = \gamma_x(\varepsilon_1, \ldots, \varepsilon k_1)$$

$$\gamma_y(\underline{\eta}) = \gamma_y(\eta_1, \ldots, \eta k_2).$$

In Equation (25) the γ's are the elementary symmetric functions of item difficulties $\varepsilon_1, \ldots, \varepsilon_{k1}$ and $\eta_1, \ldots, \eta_{k2}$ in the two Tests I and II respectively.

For each individual from Equation (25) the distribution can be derived of $a_{v.} = x$ conditional on the sum of scores across both tests $a_{v.} + b_{v.} = c_{v.} = c$.

$$p(a_{v.} = x \mid c_{v.} = c) = \frac{\gamma_x(\varepsilon)\gamma_{c-x}(\eta)\kappa_k^x}{\displaystyle\sum_{j=0}^{k1} \gamma_j(\varepsilon)\gamma_{c-j}(\eta)\kappa_v^j} \tag{26}$$

$$\kappa_v = \xi_v / \eta_v$$

Note that the abilities ξ_v and η_v (cf. Table 5) enter this distribution by their ratios $\kappa_v = \xi_v / \eta_v$. Hence, if the objective was to test for consistent abilities $H_0 : \xi_v = \eta_v$ across the two Tests I and II, the test distribution in Equation (26), being a kind of generalized hypergeometric Fisher-exact-test, will serve as the basis with $\kappa_v = 1$ $v = 1, \ldots, n$ (Allerup, 1986). Even the more relaxed hypothesis of a general "shift", Δ, in the level of abilities from Test I to Test II (viz., $H_1 : \xi_v = \Delta \eta_v$) is captured by Equation (26) by testing $\kappa_v = \Delta$ $v = 1, \ldots, n$. For these statistical tests to be carried out on real data it is necessary in case (a) to obtain access to the values of the difficulties ε_i and η_i of the two Tests I and II, either as known values from previous studies of the two reading tests, or as a result of estimating the parameters based on single-item data $((a_{vi}))$ and $((b_{vi}))$. In any case it turns out to be simple χ^2s adding up along the conditioning "c_v-diagonals" in the $k1 \times k2$ cross tabulation of $(a_{v.}, b_{v.})$, using multinomials $p(a_{v.} | c_{v.} = c)$ on each c_v-diagonal (Allerup, 1986).

If, on the other hand, H_0 and H_1 are both rejected, the alternative hypothesis remains that the subjects do have different abilities in Tests I and II. In response to the question initially raised, it is necessary, therefore to measure the magnitude of span in ability between the two tests by simply estimating the κ_vs (e.g., as $\kappa_v \approx \hat{\xi}_v / \hat{\eta}_v$ – the ratio of estimated abilities obtained from each of the Tests I and II).

At the same time, it can be seen that this "correct" model-derived estimator for κ_v is not a function of the scores $a_{v.}$ and $b_{v.}$ that is comparable to the specifications in the four methods mentioned above. In fact, the usual maximum likelihood estimates (e.g., of $\xi_1, \ldots, \xi_n$) are obtained from solving the equations

$$r = \sum_{j=0}^{k1} \frac{\varepsilon_i \xi}{1 + \varepsilon_i \xi} \tag{27}$$

$$r = 1, \ldots, k1$$

with $\varepsilon_1, \ldots, \varepsilon_{k1}$ replaced by their conditional maximum likelihood estimates; it makes only little difference if the conditional maximum likelihood estimates are applied for the ξs (Holst, 1993).

An idea of the mathematical structure of the estimator for κ_v is available, however, when trying to solve the problem of estimating the difference between Tests I and II, κ_v, in case (b), where only the scores $a_{v.}$ and $b_{v.}$ are known. Under these circumstances it is impossible, of course, to conduct proper statistical tests of H_0 or H_1, since the item difficulties $\varepsilon_1, \ldots, \varepsilon_{k1}$ and $\eta_1, \ldots, \eta_{k2}$ are all unknown, and it is necessary to rely on appropriate approximate methods.

These are, like the development from Equation (22) to Equation (24), at hand if the unconditional probability $p(a_{vi} = 1)$ of a correct response is identified with the conditional probability $p(a_{vi} = 1 | a_{v.} = r)$ given the score $a_{v.} = r$

$$p(a_{vi}=1)=\frac{\varepsilon_i\xi_v}{1+\varepsilon_i\xi_v}\approx p(a_{vi}=1\,|\,a_v=r)=\frac{\varepsilon_i\dfrac{\gamma_{r-1}^{(i)}}{\gamma_r^{(i)}}}{1+\varepsilon_i\dfrac{\gamma_{r-1}^{(i)}}{\gamma_r^{(i)}}}. \tag{28}$$

For score value $=r$, this leads to the identification of the ability $\xi(r)$, for individuals with $a_v=r$

$$\xi(r)\approx\frac{\gamma_{r-1}^{(i)}}{\gamma_r^{(i)}} \tag{29}$$

If all εs are equal $\varepsilon_1=\ldots=\varepsilon_{k1}=1$ (εs constrained by $\pi\varepsilon_i=1$) the γs are binomial coefficients

$$\gamma_r(\underline{\varepsilon})=\binom{k1}{r} \tag{30}$$

and the ratio in Equation (29) reduce to

$$\xi(r)\approx\frac{r}{k1-r} \tag{31}$$

Even if the log difficulties $\theta_i=\log(\varepsilon_i)$ $i=1,\ldots,,k1$ are uniformly spread in the normally used interval $[-2.0, 2.0]$, Equation (31) represents a good mathematical approximation to the ratio of γs in Equation (29). Together with a similar approximation for the ηs

$$\eta(r)\approx\frac{r}{k2-r} \tag{32}$$

it is possible to obtain the final approximation for $\kappa_v=\xi_v/\eta_v$. In fact, observing score value $a_v=r$ in Test I and score value $b_v=s$ in Test II Equations (31) and (32) provide the following approximate estimate of $\kappa(r)$ (i.e., κ_v values for individuals with score $=a_v=r$)

$$\kappa(r,s)\approx\frac{r/(k1-r)}{s/(k2-s)}$$

$$r=1,\ldots,k1-1 \tag{33}$$

$$s=1,\ldots,k2-1$$

By Equation (33) a solution to the initial problem has been reached, and it can be seen that the mathematical structure of the estimate of κ_v is that of a ratio between two odd values (i.e., an "odds ratio" of the number of correct responses for Tests I and II). The

Table 6 Hypothetical item difficulties $\varepsilon_{i,i}$ $i = 1, \ldots, 10$ in Tests I and II

	Item No.									
	1	2	3	4	5	6	7	8	9	10
Test I	0.67	1.79	2.01	0.22	0.67	1.12	1.56	0.45	1.79	2.01
Test II	0.27	2.72	0.54	0.82	2.45	0.54	1.90	1.09	1.36	0.82

individual measure κ_v of discrepancy between Test I and Test II abilities is obtained under the Rasch model and it is, although mathematically simple in structure, incompatible with simple differences, z-scores, etc. mentioned above. Note, that under the condition $\varepsilon_1 = \ldots = \varepsilon_{kl} = 1$ Equation (26) becomes a simple, extended hypergeometric distribution, known from analysis of 2×2 contingency tables, and the "natural" estimate of κ_v (Johnson and Kotz, 1969) is exactly the odds ratio estimate listed in Equation (33).

The power of the $\kappa(r, s)$ approximation in Equation (33) can be illustrated by an example. Taking items in Tests I and II with assumed item difficulties ε_i, η_i $i = 1, \ldots, 10$ ($k = 10$) listed in Table 6, the maximum likelihood estimates $\xi(r)$, $\eta(r)$ $r = 1, \ldots, 9$ corresponding to the nine score levels ($r = 1, \ldots, k - 1$) can be calculated from Equation (27) and, finally the true theoretical $\kappa(r, s) = \xi(r)/\eta(r)$ $r = 1, \ldots, 9$ can be determined. These values, listed in Table 7, can be compared with the approximate $\kappa(r, s)$ values listed in Table 8.

It is seen that not only does Equation (33) approximate the true, Rasch-model-derived measure κ of individual distance between Tests I and II, mathematically a structure similar to well-known terms from contingency table analysis, but it also enjoys the property of being easy to calculate by means of the test scores r, s only (see also: *Rasch Measurement Models; Sufficient Statistics in Educational Measurement*).

Table 7 True, theoretical log κ_r-values for any combination of scores, r, s, on Tests, I and II, with item difficulties $\varepsilon_{i,i}$ given in Table 6

Test I s↓	Test II (Total score $r =$)								
	2	3	4	5	6	7	8	9	10
1	− 0.01	− 0.81	− 1.37	− 1.83	− 2.26	− 2.68	− 3.13	− 3.65	− 4.34
2	0.79	− 0.01	− 0.57	− 1.03	− 1.46	− 1.89	− 2.35	− 2.89	− 3.67
3	1.34	0.53	− 0.02	− 0.49	− 0.92	− 1.35	− 1.81	− 2.37	− 3.17
4	1.79	0.99	0.44	− 0.03	− 0.46	− 0.89	− 1.36	− 1.92	− 2.73
5	2.22	1.42	0.87	0.41	− -0.03	− 0.46	− 0.93	− 1.49	− 2.31
6	2.65	1.85	1.31	0.84	0.41	− 0.02	− 0.49	− 1.06	− 1.88
7	3.10	2.32	1.78	1.32	0.89	0.45	− 0.01	− 0.58	− 1.40
8	3.63	2.87	2.34	1.89	1.46	1.03	0.56	0.00	− 0.82
9	4.34	3.66	3.16	2.71	2.29	1.86	1.39	0.83	0.01

Table 8   Log odds values based on total scores r, s as approximate $\kappa(r, s)$ values in Equation (33)

Test I s	Test II (Total score r =)								
	1	2	3	4	5	6	7	8	9
1	− 0.00	− 0.81	− 1.35	− 1.79	− 2.20	− 2.60	− 3.04	− 3.58	− 4.39
2	0.81	0.00	− 0.54	− 0.98	− 1.39	− 1.79	− 2.23	− 2.77	− 3.58
3	1.35	0.54	0.00	− 0.44	− 0.85	− 1.25	− 1.69	− 2.23	− 3.04
4	1.79	0.98	0.44	0.00	− 0.41	− 0.81	− 1.25	− 1.79	− 2.60
5	2.20	1.39	0.85	0.41	0.00	− 0.41	− 0.85	− 1.39	− 2.20
6	2.60	1.79	1.25	0.81	0.41	0.00	− 0.44	− 0.98	− 1.79
7	3.04	2.23	1.69	1.25	0.85	0.44	0.00	− 0.54	− 1.35
8	3.58	2.77	2.23	1.79	1.39	0.98	0.54	0.00	− 0.81
9	4.39	3.58	3.04	2.60	2.20	1.79	1.35	0.81	0.00

References

Allerup, P. 1985. *Why I Like to Read: Statistical Analysis of Questionnaire Data*. Danish Institute for Educational Research, Copenhagen.

Allerup, P. 1986. *Statistical Analysis of MADRS: A Rating Scale*. Danish Institute for Educational Research, Copenhagen.

Allerup, P. 1987. *Raschmodeller – Principper og Anvendelse*. Danish Institute for Educational Research, Copenhagen.

Allerup, P. 1994. Development of the reading scales. In: Beaton, A. E. (ed.) 1993. *IEA Reading Literacy Study; Technical Report*. International Association for the Evaluation of Educational Research, The Hague.

Andersen, E. B. 1973. *Conditional Inference and Models for Measuring*. Mentalhygiejnisk Forlag, Copenhagen.

Andrich, D. 1978. A rating formulation for ordered response categories. *Psychometri.* **43**(4), 561–73.

Bryant, P. and Impey, L. 1986. The similarities between normal readers and developmental and acquired dyslexics. *Cog.* **24**, 121–37.

Elley, W. B. 1992. *How in the World do Students Read? IEA Study of Reading Literacy*. International Association for the Evaluation of Educational Research, The Hague.

Fischer, G. 1974. *Einführung in die Theorie Psychologischer Tests*. Huber, Bern.

Holst, C. 1993. *Item Response Theory*. Danish Institute for Educational Research, Copenhagen.

Johnson, N. L. and Kotz, S. 1969. *Discrete Distributions*. Wiley, New York.

Lord, F. M. 1980. *Applications of Item Response Theory to Practical Testing Problems*. Erlbaum, Hillsdale, New Jersey.

Lord, F. M. and Novick, M. R. 1968. *Statistical Theories of Mental Test Scores*. Addison-Wesley, Reading, Massachusetts.

Mitterer, J. O. 1982. There are at least two kinds of poor readers: Whole-word poor readers and recoding poor readers. *Canadian Journal of Psychology* **36**(3), 445–61.

Olson, R. K. 1985. Individual and developmental differences in reading disability. In: Mackinnon, G. E. and Waller, T. G. (eds.) 1985. *Reading Research: Advances in Theory and Practice*. Academic Press, London.

Rao, C. R. 1952. *Advanced Statistical Methods in Biometric Research*. Wiley, New York.

Rasch, G. 1960. *Probabilistic Models for some Intelligence and Attainment Tests*. Danish Institute for Educational Research, Copenhagen.

Rasch, G. 1966a. An informal report on a theory of objectivity in comparisons. *Proc. of the NUFFIC Int. Summer Session in Science at "Het Oude Hof"*, The Hague.

Rasch, G. 1966b. An individualistic approach to item analysis. In: Lazarsfeld, P. F. and Henry, N. W. (eds.) 1966. *Readings in Mathematical Social Science*. MIT Press, Chicago, Illinois.

Rasch, G. 1966c. An item analysis which takes individual differences into account. *Br. J. Math. S. Psych.* **19**(1), 49–57.

Rasch, G. 1968. A mathematical theory of objectivity and its consequences for model construction. Paper presented at the European Meeting on Statistics, Econometrics and Management Science, Amsterdam.

Rasch, G. 1977. On specific objectivity – an attempt at formalizing the request for generality and validity of scientific stateements. In: *Danish Yearbook of Philosophy*. Munksgaard, Copenhagen.

Snowling, M. J. 1980. The development of grapheme–phoneme correspondence in normal and dyslexic readers. *Journal of Experimental Child Psychology* **29**(2), 294–305.

6 Rasch Measurement Models

B. D. Wright

The "Rasch measurement" models developed by Danish mathematician Georg Rasch between 1951 and 1959 and explained in his 1960 book, *Probabilistic Models for Some Intelligence and Attainment Tests*, are the most important advance in psychometrics since Thurstone's 1927 *Law of Comparative Judgment*. Objective measurement depends on measuring instruments which function independently of the objects measured. This requires a response model for calibrating their functioning which can separate the effects of instrument and object. Rasch was the first psychometrician to realize the necessity and sufficiency for objectivity of logistic response models with no interaction terms. The methods introduced in his book go far beyond measurement in education or psychology. They exemplify the principles of measurement on which all scientific objectivity is based.

Rasch models are practical realizations of "fundamental measurement". When data can be selected and organized to fit a Rasch model, the cancellation axiom of additive conjoint measurement is satisfied, a perfect Guttman order of response probabilities and hence of item and person parameters is established, and items are calibrated and persons measured on a common interval scale.

The nuclear element from which all Rasch models are built is

$$P(x; \beta, \delta) = \exp(\beta - \delta)/[1 + \exp(\beta - \delta)] \qquad (1)$$

with raw-score statistics r for person parameter β, and s for item parameter δ. The linear relation between β and δ in the exponent enables $P(x; \delta | r)$ to be noninformative concerning β, and $P(x; \beta | s)$ to be noninformative concerning δ. It follows that r is sufficient for x concerning β and ancillary concerning δ, while s is sufficient for x concerning δ and ancillary concerning β. Margining to r and s estimates β and δ sufficiently while conditioning on s and r enables their inferential separation.

1 Rasch models

The Poisson and item analysis models introduced in Rasch's 1960 book belong to the family of measurement models described by him in his 1961 article "On general laws and the meaning of measurement in psychology". Four models from this family have come into use.

The general unidimensional Rasch model can be written

$$P[k; \beta, \delta, (k), (\phi)] = \exp[\phi_x(\beta - \delta) - k_x]/\gamma \tag{2}$$

where the available response categories are labeled $0, 1, 2, \ldots, m$, a response in the xth category by a person to an item is denoted by x, the parameters β and δ are the metric positions of the person and the item on their common variable, (k) is a vector of $m + 1$ response category parameters, (ϕ) is a vector of $m + 1$ nonparametric category coefficients and

$$\gamma = \sum_{j=0}^{m} \exp[\phi_j(\beta - \delta) - k_j] \tag{3}$$

is the sum of all possible numerators.

1.1 Rating scale model

When m is finite and the $m + 1$ response categories are ordered, two simplifications occur. Andersen (1977) shows that the nonparametric category coefficients (ϕ) must be equidistant and may as well be successive integers. Andrich (1978) shows that the category parameters $(k, j = 0, m)$ can be interpreted in terms of thresholds $(\tau_j, j = 1, m)$ that govern the transitions across adjacent categories. With these interpretations

$$\phi_x = x = 0, 1, 2, \ldots, m \tag{4}$$

$$k_0 = 0 \tag{5}$$

$$k_x = \sum_{k=1}^{x} \tau_j \tag{6}$$

$$\gamma = 1 + \sum_{k=1}^{m} \exp\left[x(\beta - \delta) - \sum_{j=1}^{k} \tau_j \right] \tag{7}$$

and Rasch's general unidimensional model becomes the rating scale model studied by Andrich (1978) and Wright and Masters (1982)

$$P[x; \beta, \delta, (\tau)] = \exp\left[x(\beta - \delta) - \sum_{j=1}^{x} \tau_j \right] \Big/ \gamma \tag{8}$$

1.2 Poisson model

When the response process allows x to take any positive integer so that $m = \infty$, and individual contributions to x occur independently, then $k_x = \log(x!)$, $\gamma = \exp[\exp(\beta - \delta)]$ and the rating scale model becomes the Poisson model Rasch used for the analysis of oral misreadings and reading speeds

$$P[x; \beta, \delta] = \exp[x(\beta - \delta)]/x! \exp[\exp)\beta - \delta)] \qquad (9)$$

1.3 Partial credit model

When the thresholds τ_j are individualized to the item difficulties δ to form $\delta_j = \delta + \tau_j$ so that each item has its own set of internal step difficulties and $\delta_0 = 0$, the model becomes:

$$P[x; \beta, (\delta_j)] = \exp \sum_{j=0}^{x} (\beta - \delta_j) \bigg/ \sum_{k=0}^{m} \exp \sum_{j=0}^{k} (\beta - \delta_j) \qquad (10)$$

which is useful for the analysis of graded performance and partial credit data (Wright and Masters, 1982).

1.4 Item analysis model

When there are only two alternatives so that $m = 1$, the model becomes

$$P(x; \beta, \delta) = \exp[x(\beta - \delta)]/[1 + \exp(\beta - \delta)] \qquad (11)$$

which is the simple logistic "Rasch model" so widely used for the sample-free calibration of educational test items and the test-free measurement of individual attainment (Wright and Stone, 1979).

These four models, and a fifth for finite numbers of independent trials, can be expressed in the partial credit form as in Equation (10) by specifying

$$\delta_j = \delta_j \qquad \ldots \ldots \ldots \ldots \ldots \ldots \text{ partial credit}$$
$$\delta_j = \delta \qquad \ldots \ldots \ldots \ldots \ldots \ldots \text{.item analysis}$$
$$\delta_j = \delta + \tau_j \ldots \ldots \ldots \ldots \ldots \ldots \text{ rating scale}$$
$$\delta_j = \delta + \log(j) \ldots \ldots \ldots \ldots \ldots \text{ . Poisson counts}$$
$$\delta_j = \delta + \log(j) - \log(m + 1 - j) \ldots \ldots \text{ binomial trials}$$

2 Methods of estimation

Measurement models require methods for estimating their parameters. The LOG method Rasch used for item calibration in 1953 is easy to follow and brings out the necessity of additivity in the construction of a measuring system.

Rasch also describes a pairwise calibration in which the ability parameters of persons scoring one or two-item tests cancel when estimating the difficulty difference between the two items. This leads to a PAIR method of item calibration in which items are tabulated against one another in all possible pairs and the responses of persons attempting each pair but succeeding on only one item provide the item calibrations.

The PAIR method can also be used for person measurement because the difficulty parameters of items attempted by both of a pair of persons but succeeded on by only one of them cancel when estimating the ability difference between the two persons. In this case the persons are tabulated against one another in all possible pairs and responses to items attempted by each pair of persons, but succeeded on by only one of them, provide the person measures. This method is useful when one has too few persons to establish a useful item calibration and is sufficiently satisfied that the items work together to define a useful variable to get along without verifying this by trying to calibrate them.

LOG uses ability estimates when the ability parameters could have been removed by conditioning. PAIR does not use information based on item relationships more complex than pairwise comparisons. To improve on this, Rasch outlines a conditional method of estimation, FCON, in which all person parameters are explicitly removed by conditioning and all data are used for item calibration. Simulations carried out in 1965 and 1974 (Wright and Douglas, 1977a), however, show that tests exceeding 30 or 40 items can encounter round-off errors which spoil FCON estimates. This provoked the development of an unconditional counterpart UCON (Wright and Panchapakesan, 1969). In FCON, the person parameters are replaced by a term indexed to items as well as person scores, and calculated from symmetric functions of item estimates. In UCON this term is indexed to person scores only and its variation over items averaged out.

Rasch (1960, p. 182) shows why UCON works. The symmetric functions σ_{ri} in FCON can be written $\beta_{ri} = \log(\sigma_{r-1,i}/\sigma_{ri})$ so that the conditional probability of a person with score r succeeding on item i becomes

$$\exp(\beta_{ri} - \delta_i)/[1 + \exp(\beta_{ri} - \delta_i)] \tag{12}$$

The item parameter β_{ri}, which replaces person parameter β_r is calculated from the set of item difficulties with δ_i removed. But removing the current estimates d_i one at a time has little effect on the matrix of estimates (b_{ri}). As a result person parameter conditioning is well-approximated by reducing each vector of (b_{ri}) to b_r so that the working probability of a person with score r succeeding on item i becomes

$$\exp(b_r - d_i)/[1 + \exp(b_r - d_i)] \tag{13}$$

Wright and Douglas (1977b) show that the average effect of using b_r, the estimated ability of any person with score r, instead of calculating the symmetric functions of the item difficulty estimates, can be removed by multiplying centered UCON item difficulties by $(L - 1)/L$, where L is the number of items.

If items and persons are more or less normally distributed, an even simpler method of estimation, PROX, can be used (Wright and Douglas, 1977a; Wright and Stone, 1979, pp. 30–45). The PROX equation for estimating item difficulty d from item score s in a sample of N persons normally distributed in ability with mean M and standard deviation S is

$$d = M + [1 + (S/1.7)^2]^{1/2} \log[(N-s)/s] \qquad (14)$$

The divisor 1.7 scales the standard deviation of item difficulty from logits to probits. When persons and items are symmetrically distributed around one mode and targeted on one another, PROX produces item estimates equivalent to those of UCON or FCON.

Once an item bank is calibrated, a person can be measured with any suitable selection of items. An especially reasonable choice is a sequence of items evenly spaced over the region where the person is thought to be. This motivates an interest in how to estimate measures from tests of evenly spaced items.

The UFORM method for estimating person ability b from relative score $f = r/L$ on a uniform test of L items with average difficulty H and difficulty range W is

$$b = H + (f - 0.5)W + \log(A/B) \qquad (15)$$

where

$$A = 1 - \exp(-fW), \text{ and}$$

$$B = 1 - \exp[-(1-f)W]$$

This makes the transformation of test scores into measures simple. The ability measure implied by $f = r/L$, a proportion correct on a particular uniform test, is determined by adding H, the average difficulty of the test items and the easily tabled increment based on f and W given above. A standard error for this measure can be calculated by looking up an error coefficient in a corresponding table and dividing it by the square root of L (Wright and Stone, 1979, pp. 143–51).

3   The analysis of fit

Before estimates are used as calibrations and measurements, it is necessary to verify that the data from which they came are suitable for measuring. The requirements for measuring are specified by the model. If the data cannot be managed by the model, then they cannot be used to calibrate items or measure persons. To evaluate the fit between data and model, the validity of item response patterns must be examined during item calibration, and the validity of person response patterns examined during measurement.

The fit analysis Rasch (1960, pp. 88–105) applies to his LOG method of estimating parameters for the item analysis model is simple and elegant. Its graphical form brings out the essential part additivity plays in the construction of measures. A useful alternative for the item analysis model is to compare each response of each person to each item with its estimated expectation $p = \exp(b-d)/[1 + \exp(b-d)]$ in which b and d are the current estimates of person ability and item difficulty. When this comparison is summarized over persons for an item, it indicates the overall validity of that item. When it is summarized

over items for a person, it indicates the overall validity of that person's responses (Wright
and Stone, 1979, pp. 66–80). More detailed and more sensitive analyses of fit can be
implemented by partitioning these comparisons into relevant classes of items and/or
persons and analyzing the variance structure of these partitions.

If the observed response $x = 0$ or 1 has an expectation E estimated by

$$Ex \simeq \exp(b - d)/[1 + \exp(b - d)] = p \tag{16}$$

in which b and d are used exactly as they come from the estimation procedure (*before*
unbiasing by $(L - 1)/L$ in the case of UCON) and a variance estimated by

$$Vx \simeq p(1 - p) = w \tag{17}$$

then the BIAS statistic

$$g = \Sigma(x - p)/(wL)^{1/2} \tag{18}$$

and the NOISE statistics

$$v_1 = \Sigma[(x - p)^2/w]/L$$
$$v_2 = \Sigma(x - p)^2/\Sigma w \tag{19}$$

with expectations

$$Eg = 0$$
$$Ev = 1 \tag{20}$$

and variances

$$Vg = 1$$
$$Vv_1 = \Sigma[(1/w) - 4]/L^2 \tag{21}$$
$$Vv_2 = (\Sigma w - 4\Sigma w^2)/(\Sigma w)^2$$

test the fit of responses (x) to their corresponding expectations (p). The average restriction
in the mean squares caused by replacing the unknown probabilities by estimates based on
N persons taking L items can be corrected by multiplying v by $[NL/(N - 1)(L - 1)]$. The
development and use of fit statistics for the other models are discussed and illustrated in
detail by Wright and Masters (1982, pp. 90–117).

4 Applications of Rasch measurement

4.1 Item banking

When test items are constructed so that they calibrate along a single dimension, and when
they are used so that they retain these calibrations over a useful realm of application, then

a scientific tool of great simplicity and far-reaching potential becomes available. The resulting "bank" of calibrated items defines the variable in exquisite detail. Its item contents serve the composition of an infinite variety of pre-equated tests: short or long, easy or hard, wide in scope or sharp in focus. Neither the difficulty nor shape of these tests need have any effect on their equating. All possible scores on all possible tests are automatically equated in the measures they imply through the common calibrations of their bank items. Whatever the test, its measures are expressed on the common variable defined by the bank. Furthermore the validity of these calibrations and of each measure made with bank items can be verified at every step.

4.2   Test design

The positioning of items along the dimension they define makes test design easy. Tests can be targeted on any region along the variable represented by calibrated items. The items chosen for a particular test can be spread over the target region in whatever way is most informative. The best designs are obtained by bunching items at decision points to maximize decision information and by spreading them evenly over targets to maximize target information.

4.3   Tailored testing

The basic recipe for turning $f = r/L$, the proportion of correct answers on a test of average item difficulty H, into b, its corresponding measure in logits, is:

$$b = H + \log[f/(1 - f)] \tag{22}$$

A simple formula for optimal sequential testing follows. If each succeeding item is chosen on the basis of prior performance, the logit difficulty of the best next item can be estimated from h, the average logit difficulty of preceding items, and f, the proportion of these items answered correctly,

$$d = h + \log[f/(1 - f)] \tag{23}$$

The final measure equals the last difficulty chosen. Response validity can be checked by periodic administrations of off-target items for which the expected response is all but certain. Should invalidity emerge it can be used to revise or terminate the session.

4.4   Self-tailoring

Persons can also make their own choice of item difficulty as they go along. The items in their test can be arranged to increase in difficulty. People may choose their own starting point. If they feel strong, they may work ahead into harder items until they reach their limit. If they feel weak, they may stay with easy items. Capitalization on opportunity can be controlled by scoring persons on all items contained in the item segment their easiest and hardest item selections embrace, whether they attempt them all or not.

4.5 Response validation

The analysis of fit enables the validity of each response to be examined. This is an important step in estimating a measure from test performance. The items used will vary in their positions along the variable. This will happen when items are spread to cover the target. It will also be forced by limitations in item resources. As the simplicity and necessity of verifying response pattern validity are appreciated, items for measuring will be selected which spread out enough to facilitate the evaluation of the response patterns they stimulate.

When items vary in their difficulty, persons are expected to do better on easier items than on harder ones. Because the response model is explicit in this regard, this expectation can be formulated into an analysis of fit for any response pattern. This enables the validity of each and every test performance to be examined before any measures estimated from it are reported.

4.6 Item bias

The analysis of response pattern fit allows each person's item responses to be diagnosed in detail. If any theory is possessed that classifies items by response format, page layout, booklet location, item text, topic, or approach, then it is possible to calculate how much each person's responses are disturbed by these categories.

When a disturbance is found, it is possible to estimate the extent to which the unusual category is biased for each person. There is no other objective basis for the analysis of item or test bias. Bias estimated from groups can never satisfy the right of each individual to be fairly treated regardless of membership.

4.7 Individual diagnosis

More important is the identification of each test taker's strengths and weaknesses and the use of this diagnosis to find what he or she needs next. Most test takers are associated with programs dedicated to improving them. The justification for testing is the intention to use tests to help test takers. For this, an item content diagnosis of each test taker's response pattern is essential. Since the response residuals from the measurement model manifest all the diagnostic information the test contains, their analysis is also all that can be done statistically.

5   Connections with traditional test statistics

The person and item statistics of a Rasch analysis do not correspond directly to the indices of item difficulty, test reliability, and test validity of traditional test theory. Nevertheless, Rasch item difficulties and person abilities are closely related to traditional p-values and test scores, and the Rasch model provides valuable insight into traditional concerns for test reliability and validity.

5.1 Item p-values

The traditional approach to item difficulties uses a "*p*-value" or "proportion of persons attempting the item who are successful". These *p*-values have two shortcomings. First, they are dependent upon the abilities of the persons who took the test: the more able the persons, the higher the proportion of persons succeeding on each item. This makes it awkward to compare the difficulties of items taken by different groups of persons. Second, because they are bounded by zero and one, *p*-values cannot form an interval or linear scale: equal differences in *p*-values cannot represent equal differences in item difficulties.

Rasch item difficulty estimates are freed from both of these shortcomings. The way in which this is done can be illustrated for the particular case in which person abilities are assumed to be normally distributed. In this case, sample-free item difficulties can be approximated from item *p*-values using the formula

$$D = M + Y[\log(1-p)/p]$$

where $Y = (1 + S^2/2.89)^{1/2}$, D is the Rasch item difficulty, M is the mean ability for the sample of persons, S is the standard deviation of these abilities, and p is the traditional item *p*-value. The values of D, M and S are in logits on the linear scale shared by item calibrations and person measures and the factor $2.89 = 1.7^2$ rescales the normal distribution to follow the logistic.

This formula removes the two shortcomings of *p*-values. First, the *p*-values are transformed onto a linear scale by the logit function $\log[(1-p)/p]$. Second, this transformed *p*-value is rescaled so that the influence of the sample standard deviation S and sample mean M are removed. If the person abilities are more or less normally distributed, the resulting item calibration D is sample free. This means that the difficulties of items can be compared even though they might come from quite different samples of persons.

5.2 Test scores

A Rasch ability estimate is reported for each person taking a test, provided that the test is not so easy that the person answers all items correctly or so difficult that they are able to answer none. The traditional approach to reporting a person's ability is to count the number of correct answers made and to report either this raw test score or some norm-based transformation of it. But like *p*-values, these raw scores have two shortcomings. First, they are dependent upon the difficulties of the items in the test. If the items are easy, raw scores will be high. This makes it awkward to compare the abilities of persons taking tests of different difficulty. The second disadvantage is that, because they are bounded by zero and the maximum possible score, raw scores are also not on an interval scale. The result is that a difference of one score point does not represent the same difference in ability from one end of the score range to the other.

Rasch ability estimates are freed of these disadvantages. Under the assumption that items are normally distributed, ability estimates can be approximated from raw scores with the formula:

$$B = H + X \{ \log[r/(L - r)] \}$$

where $X = (1 + W^2/2.89)^{1/2}$, B is the person's ability measure, H is the mean difficulty of the test items, W is the standard deviation of these item difficulties, all in logits, r is the person's raw score, and L is the number of items in the test. Once again, the disadvantages of the raw scores are removed in two steps: First by transforming the scores onto an interval scale using the transformation $\log[r/(L - r)]$ and second by removing the influence of the mean and standard deviation of the test item difficulties. If the item difficulties are more or less normally distributed, then the resulting person measures are test free. This means that they can be compared even though persons take quite different sets of items. The general formula which can be used for any distribution of item difficulties is slightly more complicated. For details, see Wright and Stone (1979).

5.3 Reliability

The reliability of a test is intended to specify the accuracy with which the test measures the variable it is designed to measure. The traditional formulation of test reliability can be derived from a "true score" model which assumes that the observed test score of each person can be resolved into two components: an unknowable true score and a random error. The reliability of a test is defined as the proportion of a sample's observed score variance SD^2, which is due to the sample's true-score variance ST^2

$$R = ST^2/SD^2 = 1 - (SE^2/SD^2)$$

where $SD^2 = ST^2 + SE^2$, and SE^2 is the error variance of the test, averaged over that sample.

The size of this traditional reliability coefficient, however, depends not only upon the test-error variance SE^2 which describes how precisely the test measures (i.e., for a given ST^2, the greater the precision of measurement, the smaller SE^2, and so, the larger R), but also on the sample true-score variance ST^2 which describes the ability dispersion of the sample (i.e., for a given SE^2, the greater the sample true-score variance ST^2, the larger R). Rather than combining ST^2 and SE^2 into one compound statistic which is easily mistaken for a sample-free index of how accurately a test measures, it is more useful to distinguish between these two components of variance in the traditional reliability expression and to examine ST^2 and SE^2 separately.

The observed sample variance SD^2 can be calculated directly from the observed measures, but the test error variance SE^2 must be derived from a model describing how each score occurs. The traditional approach to estimating this error variance is to estimate the reliability first. This is done in various ways, for example, by calculating the correlation between repeated measurements under similar conditions, or by correlating split halves, or by combining item point biserials. An average error variance for the test with this particular sample is then estimated from $SD^2(1 - R')$ where R' is the estimate of R.

The magnitude of the estimated reliability R', however, also depends upon a third factor, namely the extent to which the items in the test actually work together to define one variable. The traditional estimate of R can be expressed as a function of an observed sample variance SD^2 and an "actual" test-error variance SA^2:

$$R' = 1 - (SA^2/SD^2)$$

This actual error variance has two parts. The modeled test error variance SE^2 is its theoretical basis. But it is also influenced by the extent to which the items actually fit together, and are thus internally consistent. When item inconsistency is estimated by a fit mean square V for the test as a whole, then the actual error variance is

$$SA^2 = V.SE^2$$

so that the estimated reliability becomes

$$R' = 1 - (V.SE^2)/SD^2$$

The Rasch analysis resolves these complications by dealing separately with each of the three components V, SE and SD which are submerged in the traditional test reliability coefficient.

First, the model provides a direct estimate of the modeled error variances SE^2. This modeled error indicates how precisely each person's ability can be estimated when the test items are internally consistent. Unlike the traditional reliability coefficient, this estimate is not influenced by any sample variance or fit and is not sample specific. It is a sample-free test characteristic which estimates how precisely any person's ability can be estimated from their particular score, regardless of any sample to which they may belong. Also unlike the traditional reliability coefficient, this estimate is not an average for the entire test, but is particular to whatever test score is actually obtained.

Under a Rasch analysis, the term "reliable" is best reserved for this single score-specific, sample-free aspect of traditional reliability. Rather than referring to the reliability of a whole test with respect to some sample, the term is used to describe the precision of each person's measure. Analogously, the estimate of the standard error for each item makes it possible to refer to the "reliability" of each item's calibration.

Once values for the test measurement error SE^2 of each person observed are available, an estimate of the true-sample variance ST^2 can be obtained:

$$ST^2 = SD^2 - MSE$$

where MSE is the sample mean of the individual error variances:

$$MSE = \left(\sum_{n=1}^{N} SE^2 \right) \Big/ N$$

The third factor influencing estimates of traditional test reliability, namely the internal consistency among items, is treated as the "internal validity" of the instrument.

5.4   Validity

In traditional test theory, a distinction is made between internal and external validity. The usual statistics employed to assess the internal validity of a test are the item point biserials and their accumulation into the test reliability estimate. Since the magnitude of this item statistic depends on the ability distribution of the sample, in particular, on the relationship between the item p-value and the sample ability spread, it has the disadvantage of being sample dependent. When an explicit measurement model is used, the internal validity of a test can be analyzed in terms of the statistical fit of each item to the model in a way that is independent of the sample ability distribution. A mean-square test of fit can be used to estimate the extent to which the data on each item are consistent with the latent variable implied by the collection of items in the test. The evaluation of this fit is a check on internal validity. If the fit statistic of an item is acceptable, then the item is "valid".

The pattern of each person's performances can be analyzed in the same way and, if the fit statistic for a person's performance is acceptable, then that person's test performances are interpreted as a "valid" basis for inferring a measure of that person's ability. To the extent that a person's test performances do not approximate the model (e.g., if the person tends to get easy items wrong and hard items right), the validity of that person's ability estimate is in doubt.

6   Conclusion

Rasch has devised a truly new approach to psychometric problems. He makes use of none of the classical psychometrics, but rather applies algebra anew to a probabilistic model. The probability that a person will answer an item correctly is assumed to be the product of an ability parameter pertaining only to the person and a difficulty parameter pertaining only to the item ... the ability assigned to an individual is independent of that of other members of the group and of the particular items with which he is tested; similarly for the item difficulty . . . . Thus Rasch must be credited with an outstanding contribution to one of the two central psychometric problems, the achievement of nonarbitrary measures. Rasch is concerned with a different and more rigorous kind of generalization than Cronbach, Rajaratnam, and Gleser. When his model fits, the results are independent of the sample of persons and of the particular items within some broad limits. Within these limits, generality is, one might say, complete (Loevinger, 1965, p. 151).

References

Andersen, E. B. 1977. Sufficient statistics and latent trait models. *Psychometri.* **42**, 69–81.

Andrich, D. 1978. A rating formulation for ordered response categories. *Psychometri.* **43**, 561–73.

Loevinger, J. 1965. Person and population as psychometric concepts. *Psychol. Rev.* **72**, 143–55.

Rasch, G. 1960. *Probabilistic Models for Some Intelligence and Attainment Tests.* Danmarks Paedogogiske Institut, Copenhagen. (Reprinted 1980 University of Chicago Press, Chicago).

Rasch, G. 1961. On general laws and meaning of measurement in psychology. *Proceedings of the Fourth Berkeley Symposium on Mathematical Statistics and Probability.* pp. 312–33.

Rasch, G. 1977. On specific objectivity: An attempt at formalizing the request for generality and validity of scientific statements. *Danish Yearbook of Philosophy* **14**, 58–94.

Wright, B. D. and Douglas, G. A. 1977a. Best procedures for sample-free item analysis. *Appl. Psychol. Meas.* **1**, 281–95.

Wright, B. D. and Douglas, G. A. 1977b. Conditional versus unconditional procedures for sample-free item analysis. *Educ. Psychol. Meas.* **37**, 47–60.

Wright, B. D. and Masters, G. N. 1982. *Rating Scale Analysis.* MESA Press, Chicago, Illinois.

Wright, B. D. and Panchapakesan, N. 1969. A procedure for sample-free item analysis. *Educ. Psychol. Meas.* **29**, 23–48.

Wright, B. D. and Stone, M. H. 1979. *Best Test Design.* MESA Press, Chicago, Illinois.

7 Partial Credit Model

G. N. Masters

The partial credit model is an extension of the Rasch model for dichotomously scored test data to outcomes recorded in more than two ordered response categories. One approach to the analysis of polychotomously scored data is to group the ordered response categories and to carry out multiple dichotomous analyses. A preferable approach is to implement a model for ordered response categories directly. The partial credit model is a general polychotomous item response model belonging to the Rasch family of measurement models.

1   A mathematics example

There are many situations in educational research in which students' attempts at a task can be categorized into several ordered levels of outcome. The use of multiple outcome categories is common practice when scoring performances on complex tasks like essay writing and problem-solving. But even in situations in which it is usual to score students' performances dichotomously (right/wrong), it is often possible to identify among students' "incorrect" answers varying degrees of partial understanding and so to define more than two levels of outcome on an item. This can be illustrated with the following item from a test of basic mathematics:

> A calculator shows the figure 25.634817
> Express this correct to two decimal places.

Students give a variety of answers to this item, but by far the most common are 25.63, 25.64, 2563.4817 and 0.25634817.

The usual dichotomous scoring of this item would give full credit for the first of these answers and no credit for any other. However, the second answer, 25.64, shows partial understanding: students who give this answer understand that correcting a number to two decimal places involves reducing to two the number of digits after the decimal point. These students appear to believe that because the original number is greater than 25.63 it must be rounded *up* to 25.64. The last two answers indicate no understanding of rounding and result from moving the decimal point two places (as in multiplication and division by

100). The most that can be said for these two answers is that they show some understanding of "two decimal places". This is more than can be said for the other answers that students give to this item (e.g., "25.634.817").

The answers given to this mathematics item by a group of 570-ninth grade students are summarized in Figure 1. Students' answers have been grouped to form four ordered outcome categories: 25.63; 25.64; either 2563.4817 or 0.25634817; and some other answer. The 570 students have been divided into 10 equal-sized groups on the basis of their total mathematics test scores. Students with the lowest test scores are at the bottom of Figure 1. Students with the highest test scores are at the top. Figure 1 shows the proportion of students in each test score group in each of the four outcome categories.

Among the lowest-scoring group of students (at the bottom of Figure 1), only about 20 per cent of students gave the correct answer, 25.63, to this item. The most common answer given by this group was either 2563.4817 or 0.25634817. More than 60 per cent of incorrect answers given by this low-scoring group were of this type. Among the highest-scoring group (at the top of Figure 1), about 80 per cent of students gave the correct answer. Ninety per cent of high-scoring students who gave incorrect answers to this item

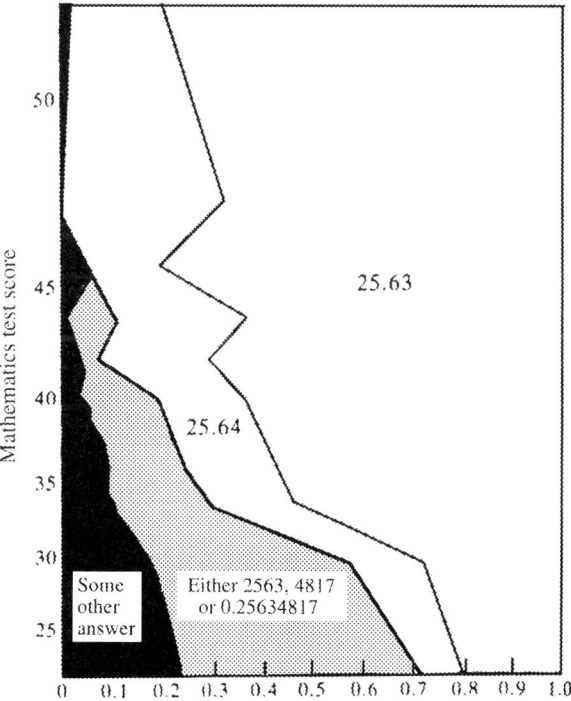

Figure 1   Observed proportions in outcome categories.

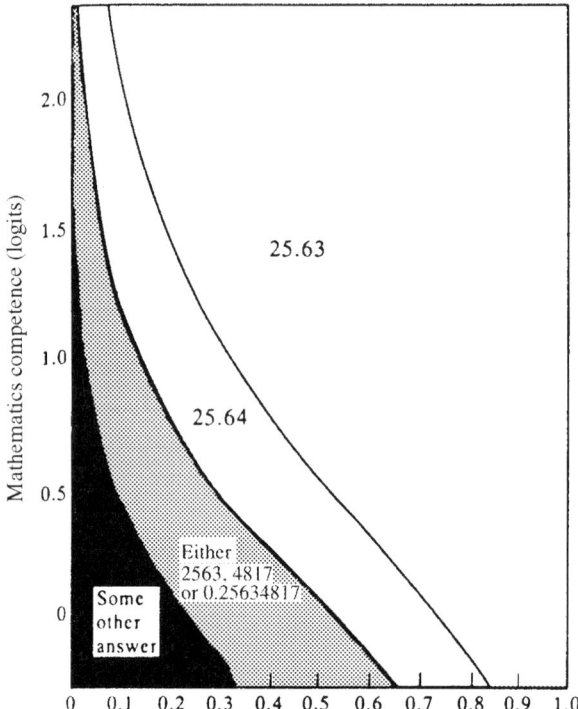

Figure 2   Modeled proportions in outcome categories.

gave the answer 25.64. Figure 1 shows that the types of errors made on this item change with increasing mathematics test score. Very few of the incorrect answers given by students with low test scores show any understanding of rounding. In fact, about 20 per cent of low-scoring students give "other" answers like 25.634.817, suggesting that these students may not even understand "two decimal places". In contrast, the incorrect answers given by high-scoring students display some understanding of rounding but reveal confusion about when to round up or down. In an instructional setting, it would be inappropriate to treat every student giving an "incorrect" answer to this mathematics item in the same way. The type of instruction in rounding decimal numbers required by most low-scoring students in Figure 1 is likely to be very different from the instruction required by most high-scoring students.

The partial credit model is a statistical model for the analysis of test and questionnaire items for which two or more ordered levels of outcome are defined. Its purpose is to model changes in the distribution of students' answers over the available outcome categories with increasing competence. For the mathematics item described above, the four outcome regions as modeled by the partial credit model are shown in Figure 2. The four regions of this map correspond to the four regions in Figure 1. The difference is that these regions no longer show the observed proportions of students in each category, but show *modeled* proportions. The basic shapes of the smooth curves in Figure 2 are fixed by the algebra of the partial credit model. The locations of these curves were estimated from the answers this group of 570 students gave to this item.

When students' answers to an item approximate the partial credit model (i.e., Figure 1 resembles Figure 2), that item can be used to help estimate students' locations on the path of developing competence that runs up the left edge of these figures. It is in this sense that the partial credit model is a "measurement" model: it provides a probabilistic connection between the categories of observed outcome on an item and locations on a latent path of developing competence. This probabilistic connection provides a basis for constructing measures of competence from students' performance on a set of items with multiple outcome categories.

2   Algebra of the model

In common with all latent trait models, the partial credit model represents each student's level of competence or achievement as a location on a continuum of increasing competence. In Figures 1 and 2, this continuum runs up the left edge of the figure. The location β_n of student n on this continuum is estimated from that student's answers to a set of appropriate items. Answers to each item are classified into a set of ordered outcome categories labeled $0, 1, 2, \ldots, m_i$ for that item. Under the partial credit model, the probability of student n's answer being in outcome category x of item i is given by

$$P_{nix} = \frac{1}{1 + \sum_{k=1}^{mi} \exp \sum_{j=1}^{k} (\beta_n - \delta_{ij})} \qquad \text{for } x = 0$$

and

$$P_{nix} = \frac{\exp \sum_{j=1}^{x} (\beta_n - \delta_{ij})}{1 + \sum_{k=1}^{mi} \exp \sum_{j=1}^{k} (\beta_n - \delta_{ij})} \qquad \text{for } x = 1, 2, \ldots, m_i \qquad (1)$$

where the parameters $\delta_{i1}, \delta_{i2}, \ldots, \delta_{im_i}$ are a set of parameters associated with item i which jointly locate the model probability curves for that item (see Figure 2). There are m_i item parameters for an item with $m_i + 1$ outcome categories.

For the mathematics item described above, four outcome categories were defined meaning that $4 - 1 = 3$ parameters δ_{i1}, δ_{i2} and δ_{i3} were estimated for this item. These three estimates and the algebra of the model provide the modeled outcome regions in Figure 2.

At any estimated level of competence β_n, the partial credit model provides the widths P_{ni0}, P_{ni1}, P_{ni2} and P_{ni3} of the four outcome regions at that level. These widths can be interpreted either as the estimated probabilities of a student at that level of competence responding in outcome categories 0, 1, 2 and 3, or as the expected proportions of students at that level of competence responding in these four categories. For values of β_n near the bottom of Figure 2, P_{ni0} and P_{ni1} are larger than P_{ni2} and P_{ni3}. For values of β_n near the top of Figure 2, P_{ni3} is large, and all other probabilities are small. A more complete discussion of the algebra of the partial credit model is provided by Masters (1982, 1988), Wright and Masters (1982), and Masters and Wright (1984).

3   Ordered outcome categories

The partial credit model could be applied to any set of test or questionnaire data collected for the purposes of measuring students' abilities, achievements, or attitudes provided that responses to each test or questionnaire item are scored in two or more ordered categories. There are many different ways in which a set of ordered outcome categories might be defined for a task. Some of these are considered below.

3.1   Levels of partial understanding

The four outcome categories defined for the mathematics item above were the product of a careful study of all answers given by students to this item. This approach to developing a set of outcome categories is described in some detail by Dahlgren (1984). For some tasks, the types of misconceptions and errors that are likely to occur will be well understood, making it possible to construct a set of outcome categories before the task is given to a group of students. For most tasks, however, the construction of a set of categories which capture levels of partial understanding will probably require a close study of students' responses. These might then be grouped according to the levels of understanding that they reflect.

> Starting with a comparatively large number of categories the researcher will gradually refine these, arriving at a smaller set of categories that may finally be difficult or impossible to collapse further (Dahlgren, 1984, p. 26).

Dahlgren describes this approach to constructing a set of outcome categories as the partitioning of the "outcome space" associated with a task. The final set of categories is then used in future applications of the task.

3.2   Multistep problems

For complex problems which require the completion of a number of steps, it is usual to identify several intermediate stages in the solution of each problem and to award partial credit on the basis of the number of steps a student completes. This scoring procedure is common in subject areas like mathematics and the physical sciences where students must first identify the problem type, select an appropriate solution strategy, and then apply this strategy which may itself involve a number of steps. By awarding credit for the steps a student has successfully completed, a set of ordered outcome categories can be defined for each multistep problem.

4   Rating scales

Another common method for recording performances on an item is to rate students' attempts at the item on a scale (e.g., 1 to 5). This scoring procedure is popular for recording performances on tasks like building a model, assembling a piece of apparatus, carrying out a procedure, and writing an essay. To ensure a degree of comparability across raters and over time, the criteria to be applied in rating performances on a task might be made explicit and accompanied by samples of student attempts at that task to illustrate the available score points.

Rating scales are also common methods of measuring attitudes and personalities. In these contexts, respondents are usually provided with a fixed set of response alternatives like "never", "sometimes", "often", "always", or "strongly disagree", "disagree", "agree", "strongly agree" to be used with all items on the questionnaire. Questionnaires of this form can be analyzed with the partial credit model. However, the fact that the response alternatives are defined in the same way for all items introduces the possibility of simplifying the partial credit model by assuming that, in questionnaires of this type, the pattern of modeled outcome regions (Figure 2) will be the same for all items on the questionnaire and that the only difference between items will be a difference in location on the measurement variable (e.g., difficulty of endorsement). This assumption yields the rating scale model (see *Rating Scale Analysis*).

4.1   Question clusters

Occasionally, test and questionnaire items come in clusters with all items in a cluster relating to the same piece of introductory text. Each item in a cluster could be treated as an independent item to be scored right or wrong. However, if items of this type are to be treated as independent dichotomously-scored items, then the assumption of local independence must be made. Each student's response to any one item must be assumed to be without the influence of his or her responses to the other items in that cluster. In most dichotomously scored tests, this is a reasonable assumption. But in an item cluster, items have a shared dependence on a common stem and so are less likely to be locally independent. In this context, it is often more appropriate to treat a cluster as a single "item" on which students' scores are counts of the questions in that cluster answered correctly and to take values between 0 and m_i (where m_i is the number of questions in the cluster). In this way, $m_i + 1$ ordered levels of outcome are defined for each cluster (Andrich, 1982; Masters and Evans, 1986).

4.2   Interactive items

Finally, ordered outcome categories can be constructed from students' performances on computer-administered items which provide feedback to students during a test. The feedback given during a test may simply inform students of their success or failure on an item and offer a second attempt if the item is failed. Failure on a second attempt might be followed by a third or fourth attempt and credit awarded on the basis of the number of attempts required to provide the correct answer. This procedure is usually referred to as "answer-until-correct" scoring. Alternatively, students failing on their first attempt at an item might be given a "hint" and offered an opportunity to try again (Trismen, 1981). Failure after a hint might be followed by further assistance and each student's score based on the number of hints required to arrive at the correct answer. This format not only defines several ordered levels of outcome for each item but also, through the careful construction of hints, might be used to trace students' misunderstandings to their source.

5   Related models

The partial credit model is a latent trait (or item response) model and, in particular, is a member of the Rasch family of latent trait models. The relationship of the partial credit model to a number of other members of this family (e.g., Poisson counts model, binomial trials model) is described by Masters and Wright (1984). Several of these related models are considered briefly below.

5.1   Dichotomous Rasch model

The dichotomous model (see *Rasch Measurement Theory*) is designed for the analysis of test items for which only two levels of outcome are defined ($x = 0$ and $x = 1$). The dichotomous model is obtained by setting $m_i = 1$ in Equation (1). This provides the model probabilities

$$P_{nix} = \frac{1}{1 + \exp(\beta_n - \delta_{i1})} \qquad \text{for } x = 0$$

$$= \frac{\exp(\beta_n - \delta_{i1})}{1 + \exp(\beta_n - \delta_{i1})} \qquad \text{for } x = 1$$

The resulting outcome map (Figure 2) contains only two regions ("fail" and "pass") and the single parameter estimate δ_{i1} locates the modeled boundary between these two regions. This model is the best known of the item response models and is widely used for the analysis of educational tests.

5.2   Rating scale model

The rating scale model (see *Rating Scale Analysis*) can be used to analyze questionnaires in which a fixed set of response alternatives like "strongly disagree", "disagree", "agree" and "strongly agree" is used with every item on the questionnaire. The rating scale model is obtained by resolving the general item parameter δ_{ij} in Equation (1) into two components: one for item i, and one associated with the transition between response alternatives $j - 1$ and j:

$$\delta_{ij} = \delta_i + \tau_j$$

The rating scale model is obtained by substituting $(\delta_i + \tau_j)$ for δ_{ij} in Equation (1):

$$P_{nix} = \frac{1}{1 + \sum_{k=1}^{m} \exp \sum_{j=1}^{k} (\beta_n - \delta_i - \tau_j)} \qquad \text{for } x = 0$$

$$= \frac{\sum_{j=1}^{x} (\beta_n - \delta_i - \tau_j)}{1 + \sum_{k=1}^{m} \exp \sum_{j=1}^{k} (\beta_n - \delta_i - \tau_j)} \qquad \text{for } x = 1, 2, \ldots, m$$

When this model is applied, a single location δ_i is estimated for each item and m parameters $\tau_1, \tau_2, \ldots, \tau_m$ are estimated for the $m + 1$ response alternatives provided with the questionnaire.

5.3   Ordered partition model

The ordered partition model (Wilson, 1993) can be applied when responses to an item are recorded in K_i categories, but some of these K_i categories, while qualitatively different, represent the same *level* of performance (i.e., the number of distinguishable levels of performance on item i, L_i, is less than the total number of response categories K_i). When $L_i = K_i$, the ordered partition model is simply the partial credit model.

The ordered partition model is obtained by introducing a vector B_i which indicates the level to which each response category belongs. If, for item i, response category k is to be assigned to level j, then $B_i(k) = j$ where j is an integer between 0 and L_i. Under the ordered partition model, the probability of person n responding in category k of item i is

$$P_{nik} = \frac{\exp(\beta_n B_i(k) + \xi_{ik})}{\sum_{h=1}^{ki} \exp(\beta_n B_i(h) + \xi_{ik})} \qquad k = 1, 2, \ldots, K_i$$

where ξ_{i1} is defined as 0 and the item parameters in the partial credit model ((δ)) can be expressed in terms of the item parameters in the ordered partition model ((ξ)). The estimation of the parameters in the model is described by Wilson and Adams (1993).

5.4   Other constraints

Other cases of the partial credit model can be generated by imposing constraints on the values of the item parameters $\delta_{i1}, \delta_{i2}, \ldots, \delta_{imi}$ for each item. One simple constraint is to restrict these parameters to a uniform spacing such that $(\delta_{i2} - \delta_{i1}) = (\delta_{i3} - \delta_{i2}) = \ldots = (\delta_{im} - \delta_{im-1}) = \sigma_i$.

Under this constraint (Andrich, 1982), only the mean item parameter δ_i and the uniform spacing σ_i are estimated for each item i. If there is a reason to expect that the outcome categories for every item on a test will be uniformly spaced, and the data conform to this expectation, this case of the partial credit model offers a more parsimonious representation than the full-rank model in that it requires the estimation of fewer parameters. This constrained version of the model may also be useful with small data sets which provide insufficient data to estimate reliably all parameters for an item.

Further cases of the partial credit model have been proposed by introducing other constraints on the item parameters (e.g., steadily increasing or steadily decreasing differences $(\delta_{i2} - \delta_{i1}) < (\delta_{i3} - \delta_{i2}) < \ldots < (\delta_{im} - \delta_{im-1})$. In general, constraints such as these are only likely to be of value if they have a basis in theory (i.e., if they follow from the way in which the ordered categories have been defined).

6   Applications

Estimation algorithms for the partial credit model are described by Masters (1982), Wright and Masters (1982), Glas and Verhelst (1989), and Wilson and Adams (1993). Computer programs to implement these algorithms have been developed by Adams and Khoo (1993) and Wright and Linacre (1992).

In 1989 a special issue of *Applied Measurement in Education* described a range of applications of the partial credit model. Some areas of application are summarized briefly below.

6.1   Variable definition

Figures 1 and 2 illustrate how, by classifying "incorrect" answers to an item into a number of ordered levels of understanding or completion, it is possible to build a more detailed picture of how competence in a subject area develops. This is an important general application of the partial credit model. The probabilistic connection between categories of observed outcome on an item and the latent continuum that these items are constructed to measure, enables each level of competence on the measurement variable to be interpreted in terms of the types of misconceptions or processing errors that are likely to be found among students at that level. Students with estimated locations near the top of Figure 2, for example, are likely to have very different misunderstandings from students with estimated locations near the bottom of this figure.

Adams *et al.* (1987) have used this method to build a detailed picture of a path of developing competence in second language learning. The items in their instrument were questions posed to second language learners in face-to-face interviews. Each learner's response to a question was rated using a set of ordered outcome categories specific to that question.

The Ministry of Education in Western Australia has taken a similar approach to analyzing students' performances on written expression tasks. They have identified a number of aspects of writing competence and have developed a set of rating points for each of these aspects of writing. Each set of ordered rating points is illustrated using samples of student writing. In this way, a number of "ladders" of developing competence corresponding to different aspects of writing ability have been constructed and calibrated. These are used as a framework for scoring students' performances on writing tasks and provide a detailed picture of the development of writing competence.

Other examples of the use of partial credit analysis in variable construction are provided by Adams *et al.* (1990), Julian and Wright (1988), and Wright and Masters (1982).

6.2   *Item banking*

Calibrated item banks are usually limited to dichotomously-scored test items. This is a serious limitation if the bank is to be used as part of a program of educational assessment. A large proportion of what is taught in schools is not adequately assessed with items that can be scored either right or wrong. If an item bank is to be useful as an assessment resource, it must be capable of incorporating calibrated tasks like essay-writing, problem-solving and model-building.

The partial credit model provides a basis for the calibration of a range of tasks which cannot adequately be scored dichotomously. If these tasks are to be calibrated and included in an item bank, then it will usually be necessary to provide explicit guides to the scoring of individual tasks, possibly with samples of student responses to illustrate the score points to be used with each task. Some experimental work on the construction of banks of nondichotomously scored items is described by Masters (1984) and Masters and Evans (1986) (see *Item Banking*).

6.3   *Computer adaptive testing*

The availability of a bank of calibrated items introduces the possibility of selecting items to suit an individual's current level of competence. If items are administered by computer, then the items to be presented to a student can be selected automatically during the course of a test. After each item is answered, the student's level of competence is re-estimated and the bank is searched for the most appropriate remaining item. This is the item that provides most information at the student's current estimate.

Computer adaptive testing can be generalized to items which use systems of partial credit scoring, thereby enabling the construction of tailored tests based on more complex outcome spaces than right and wrong answers. The simplest adaptive testing algorithm for the partial credit model uses the statistical "information" I_{ni} available from bank item i at competence level β_n. This can be calculated as

$$I_{ni} = \sum_{k=1}^{mi} (k^2 P_{nik}) - \left(\sum_{k=1}^{mi} k P_{nik} \right)^2$$

where $P_{nik}(k=1, 2, \ldots, m_i)$ is the model probability of person n with an estimated level of competence β_n giving an answer in outcome category k of item i. The value of this information might be calculated for each item in a bank, given student n's current estimate, and the item with the largest value of I_{ni} chosen as the next item to be administered to person n.

Important foundational work on the extension of computer adaptive testing procedures to items which use systems of partial credit scoring has been carried out by Koch and Dodd (1986, 1989) and Dodd and Koch (1986, 1987). They describe a number of potential applications of this methodology, including the possibility of constructing computer adaptive questionnaires in which items designed to measure attitudes or opinions might be calibrated and selected to maximize the information available from a questionnaire. Another very promising application of this method is the construction of computer adaptive tests in which feedback is provided and multiple attempts are permitted at individual computer-administered test items (see *Adaptive Testing*) (see also: *Rasch Measurement Theory; Latent Trait Measurement Models; Rasch Measurement Models; Computerized Educational Testing*).

References

Adams, R. J., Doig, B. A. and Rosier, M. 1990. *Science Learning in Victorian Schools*. Australian Council for Educational Research, Melbourne.
Adams, R. J., Griffin, P. E. and Martin, L. 1987. A latent trait method for measuring a dimension in second language proficiency. *Language Testing* 4(1), 9–27.
Adams, R. J. and Khoo, S. T. 1993. QUEST: *The Interactive Test Analysis System*. Australian Council for Educational Research, Melbourne.
Andrich, D. 1982. An extension of the Rasch model for ratings providing both location and dispersion parameters. *Psychometri.* 47(1), 105–13.
Dahlgren, L. O. 1984. Outcomes of learning. In: Marton, F., Hounsell, D. and Entwistle, N. (eds.) 1984. *The Experience of Learning*. Scottish Academic Press, Edinburgh.
Dodd, B. and Koch, W. 1986. Relative efficiency analyses for the partial credit model. Paper presented at the annual meeting of the American Educational Research Association, San Francisco, California.
Dodd, B. G. and Koch, W. R. 1987. Effects of variation in item step values on item and test information in the partial credit model. *Appl. Psych. Meas.* 11, 371–84.
Glas, C. A. W. and Verhelst, N. D. 1989. Extensions of the partial credit model. *Psychometri.* 54(4), 635–59.
Julian, E. R. and Wright, B. D. 1988. Using computerized patient simulations to measure the clinical competence of physicians. *Appl. Meas. Ed.* 1(4), 299–318.
Koch, W. R. and Dodd, B. G. 1986. Operational characteristics of adaptive testing procedures using partial credit scoring. Paper presented at the annual meeting of the American Educational Research Association, San Fransisco, California.
Koch, W. R. and Dodd, B. G. 1989. An investigation of procedures for computerized adaptive testing using the partial credit scoring. *Appl. Meas. Ed.* 2(4), 335–57.
Masters, G. N. 1982. A Rasch model for partial credit scoring. *Psychometri.* 47(2), 149–74.
Masters, G. N. 1984. Constructing an item bank using partial credit scoring. *J. Ed. Meas.* 21(1), 19–32.
Masters, G. N. 1988. The analysis of partial credit scoring. *Appl. Meas. Ed.* 1(4), 279–98.

Masters, G. N. and Evans, J. 1986. Banking non-dichotomously scored items. *Appl. Psychol. Meas.* **10**(4), 355–67.

Masters, G. N. and Wright, B. D. 1984. The essential process in a family of measurement models. *Psychometri.* **49**(4), 529–44.

Trismen, D. M. 1981. The development and administration of a set of mathematics items with hints (ETS-RB-81-5). Educational Testing Service, Princeton, New Jersey.

Wilson, M. R. 1993. The ordered partition model: An extension of the partial credit model. *Appl. Psych. Meas.*

Wilson, M. R. and Adams, R. J. 1993. Marginal maximum likelihood estimation for the ordered partition model. *J. Ed. Stat.* **18**(1), 69–90.

Wright, B. D. and Linacre, J. M. 1992. FACETS computer program. MESA Psychometric Laboratory, University of Chicago, Chicago, Illinois.

Wright, B. D. and Masters, G. N. 1982. *Rating Scale Analysis*. MESA Press, Chicago, Illinois.

Further reading

Adams, R. J. 1988. Applying the partial credit model to educational diagnosis. *Appl. Meas. Ed.* **1**(4), 347–62.

Adams, R. J. and Griffin, P. E. 1986. *Scaling Tests of Spoken Language with the Rasch Partial Credit Model.* Victorian Ministry of Education, Melbourne.

Harris, J., Laan, S. and Mossenson, L. 1988. Applying partial credit analysis to the construction of narrative writing tests. *Appl. Meas. Ed.* **1**(4), 335–46.

Masters, G. N. and Mislevy, R. J. 1993. New views of student learning: Implications for educational measurement. In: Fredericksen, N., Mislevy, R. J. and Bejar, I. I. (eds.) 1993. *Test Theory for a New Generation of Tests.* Erlbaum, Hillsdale, New Jersey. pp. 219–242.

Muraki, E. 1992. A generalized partial credit model: Application of an EM algorithm. *Appl. Psych. Meas.* **16**(2), 159–76.

Pollitt, A. and Hutchinson, C. 1987. Calibrating graded assessments: Rasch partial credit analysis of performance in writing. *Language Testing* **4**(1), 72–92.

Wilson, M. and Iventosch, L. 1988. Using the partial credit model to investigate responses to structured subtests. *Appl. Meas. Ed.* **1**(4), 319–34.

Wilson, M. R. and Masters, G. N. 1993. The partial credit model and null categories. *Psychometri.* **58**(1), 87–99.

8 Rating Scale Analysis

D. Andrich

Rating scales are used to help identify the degrees of a *property* or *trait* of an object or person when no instrument for measuring the trait is available. Dawes (1972) estimated that some 60 percent of social science studies have rated variables as the only form of dependent variable. Rating scales are used in attitude questionnaires where responses to statements take the form Strongly Disagree, Disagree, Neutral or Undecided, Agree and Strongly Agree, and in performance ratings where judges classify performances in categories like Poor, Fair, Good and Excellent. The former, with semantically opposite extremes, is a bipolar scale, while the latter is a unipolar scale. Many variants on these formats exist. In some cases, descriptions of the two extreme categories are given, with the central categories denoted by cutoff points; in others, such as judgment of performance, operational examples at each level are provided. This entry considers the procedures employed in the analysis of rating scale data.

1 Contingency tables

In many contexts, individuals who belong to defined classes are rated into categories. For example, educators involved with different aspects of education may provide an opinion on public examinations or competency testing. Table 1 shows the kind of format, called a contingency table, in which responses would be presented. A topic such as public examinations may prompt responses to more than one statement. Results are then often reported statement by statement and inferences are drawn regarding the level of support enjoyed by each issue.

2 Individual classifications

Often classifications more refined than contingency tables are required. First, the performance of each person rated, or providing an opinion, may need to be considered individually rather than as a member of some population. Second, it may be necessary to obtain more precise information about that person than can be obtained from one

Table 1 Format for ratings in contingency tables[a]
Please respond to the following statement in one of the categories provided:
There should be publicly defined standards which all students should pass before leaving high school

	Response score	Strongly disagree (0)–	Disagree (1)	Agree (2)	Strongly agree (3)	
						Total No
	Elementary	f_{10}	f_{11}	f_{12}	f_{13}	N_1
Teaching level	Secondary	f_{20}	f_{21}	f_{22}	f_{23}	N_2
	Tertiary	f_{30}	f_{31}	f_{32}	f_{33}	N_3

[a] f denotes frequency

statement or rating. Therefore, more than one statement on a related topic, or more than one task, is required. In situations with two or more statements or two or more tasks, the information is collapsed into a single value for each person. Table 2 shows the kind of format in which the responses would be presented.

Where many statements or tasks are provided, whether in opinion, attitude, performance, or achievement ratings, the statements or tasks have the same role as they do in Thurstone scales to which dichotomous responses rather than ratings are required (see Thurstone, 1927). That is, they serve to define a continuum, and the ratings can be seen as extensions and refinements to dichotomous responses such as Disagree or Agree and Correct or Incorrect. Viewed from this perspective, the increase in the number of categories beyond two helps to increase the precision. The greater the number of categories, and to the degree that the categories can be used meaningfully, the greater the precision.

Unlike performance ratings, where a rater rates a performance explicitly, the rater in attitude testing is the person whose attitude is to be assessed. Attitude questionnaires

Table 2 Format for ratings of individuals statements or tasks

Rating score x		1 0 1 2	2 0 1 2	i 0 1 2	J 0 1 2
Person (ratee)	1	x_{11}	x_{12}	x_{1i}	x_{1J}
	2	x_{21}	x_{22}	x_{2i}	x_{2J}
	3	x_{31}	x_{32}	x_{3i}	x_{3J}
	.	.	.	.	.
	.	.	.	.	.
	n	x_{n1}	x_{n2}	x_{ni}	x_{nJ}
	.	.	.	.	.
	.	.	.	.	.
	N	x_{N1}	x_{N2}	x_{Ni}	x_{NJ}

requiring such ratings are said to be of the Likert style, following the work by Likert (1932) on attitude measurement.

3   Scoring the response categories

The scoring of the ordered response categories in rating scales has received a great deal of attention. The most elementary approach follows closely the measurement analogy.

With a formalized measuring instrument, any object can be placed between the two cutoff points or thresholds on a continuum mapped onto a real line. On a typical measuring instrument the thresholds are represented by line segments of equal width which cut the real line at equal intervals. The measure then is the number of thresholds the object is seen to pass and this measure may be refined by having smaller intervals between thresholds and by having thresholds represented by finer lines. Often, measurement errors are considered sufficiently small relative to the variation of the measured property that they are ignored and the measures are then treated as continuous variables.

By analogy, in the rating scale the thresholds are placed so that they indicate equal spacing, and the raters are supposed to place their response between two thresholds. Elementary quantification and analysis extend this measurement analogy. Thus the successive categories are scored with successive integers, and the resultant numbers are again treated as continuous variables. The integers may start with either 0 or 1. For convenience here, they will be taken to commence with 0 and to have a maximum of m, where there are m thresholds and therefore $m + 1$ categories.

In the contingency table context, the relative status of each *group g* is calculated then simply by

$$r_g = \sum_{x=0}^{m} x f_{gx}$$

while in the assessment of individuals on statements or tasks i, $i = 1, \ldots, k$, the status of each *individual n* is calculated simply by

$$r_n = \sum_i x_{ni}.$$

That is, the integer ratings are simply summed. Standard analyses based on these summary statistics, and following the true score model of traditional test theory, have been developed. Guildford (1954, Chapter 11) provides a comprehensive discussion of the construction of rating scales and on the analyses using the above scoring. However, the assumption of equality of intervals, on which the integer scoring is supposed to be based, has often been questioned. This has led to more formal representations of the rating process.

4   Mathematical response models

Two qualifications to the elementary analogy to a measuring instrument are made. First, it is accepted that there is uncertainty in the classification, second that the distances between thresholds may not be equal and require estimating.

4.1   The traditional threshold model

The traditional model, originating with Thurstone (1927), assumes a single response process centered at the location μ (of a group or individual). This process may be either the cumulative normal or the logistic distribution, but because the two are indistinguishable numerically after a linear scaling, and because the latter is more tractable, it is now usually preferred. Then the probability of a response above each threshold is the area beyond the threshold in the distribution, Figure 1 shows the process for the logistic function, where the probability p_x^* of a response *above* threshold τ_x, $x = 1, m$, is given by

$$p_x^* = \int_{\tau_x}^{\infty} \frac{\exp(x - \mu)/\sigma}{\{1 + \exp(x - \mu)/\sigma\}^2} \, dx = \frac{1}{\gamma} \exp\{(\mu - \tau_x)/\sigma\} = \frac{1}{\gamma} \exp\{\alpha(\mu - \tau_x)\} \qquad (1)$$

where (i) $\alpha = 1/\sigma$ is termed the discrimination and (ii) $\gamma = 1 + \exp\{\alpha(\mu - \tau_x)\}$ which ensures that p_x^* and its complement, the probability of a response below threshold x, sum to 1. The probability p_x of a response in category x, $x = 0, 1, 2, \ldots, m$ is given simply by the difference between successive cumulative probabilities: $p_x = p_{x-1}^* - p_x^*$ with $p_0^* \equiv 1$. Note that the ratio of p_x^* and $1 - p_x^*$ gives $p_x^*/(1 - p_x^*) = \exp\{\alpha(\mu - \tau_x)\}$ of which the natural logarithm $\log p_x^*/(1 - p_x^*) = \alpha(\mu - \tau_x)$ is called the *logit*.

The parameters α, μ and τ_x may be qualified according to the context. For example, in a contingency table, the logit in Equation (1) may take the form $(\mu_g - \tau_x)$ where only the

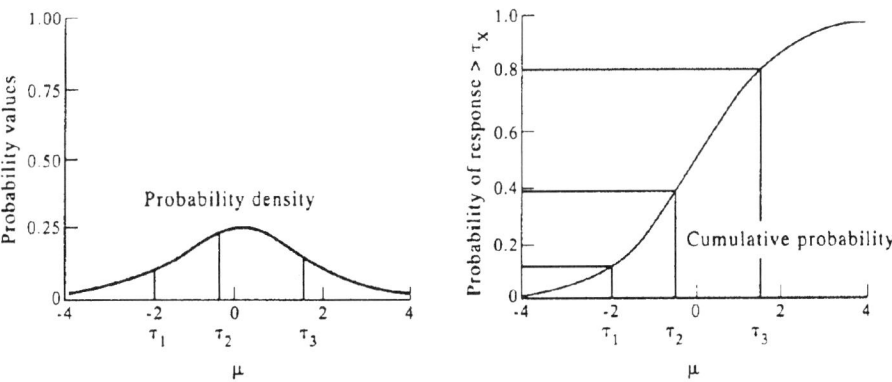

Figure 1   Probability of the response process with the Thurstone cumulative probability model.

location of the groups vary, or $\alpha_g(\mu_g - \tau_g)$, where the discrimination among groups also varies, or $\alpha_g(\mu_g - \tau_{xg})$ where the thresholds also vary among groups.

In the case for the assessment of individual n across more than one task or statement i, μ is resolved according to $\mu_{ni} = \beta_n - \delta_i$. Then if each task or statement is assumed to have the same discrimination and equal thresholds, the logit in Equation (1) may take the form $(\beta_n - \delta_i - \tau_{xi})$.

The estimation of the parameters may be carried out in various ways. In the early work, and in the case where ratings are associated with a group or population, the proportions of responses in the respective categories were taken as a direct estimate $\hat{p}_x$ of the corresponding probabilities p_x. Then the estimates of $\mu - \tau_x$ were given simply by either the corresponding standard normal deviate, or the logit given by $\log(\hat{p}_x^*/(1 - \hat{p}_x^*))$ for each group. More recent techniques involve the maximum likelihood estimation (MLE) procedure. This requires identifying the values of the parameters of the chosen model which maximize the probability for the data observed.

Likert originally correlated the estimates of locations using weights for categories in the manner described above, with those obtained by simply summing the integer scores. These correlations were generally very close to 1.00. As a result, and for simplicity, Likert and the majority of users of rating scales since then have used the simple integer scoring. That is, they have continued with the original measurement analogy.

While the characterization or measurement of individuals is often the main task, understanding and controlling the rating mechanism is also important. Therefore, researchers have continued to show concern about the assumption of equal intervals on the rating scales. The traditional threshold model and its mathematical formulation described above is one attempt to accommodate these concerns, for which Bock (1975, Chapter 8) provides the mathematics for contingency tables and Samejima (1969) for the assessment of individuals.

4.2   The Rasch rating model

A more recent formulation of a mathematical model for ratings (Andersen, 1977; Andrich, 1978) accommodates not only the features of a random response process and the estimation of thresholds, but also the simple integer scoring of the successive categories and the simple summing among tasks or statements. In addition, it has major epistemological differences with Thurstone's formulation referred to later. If p_x is again the probability that a rating, governed by a true value μ, is in category x, the model takes the form

$$p_x = \frac{1}{\gamma} \exp\left\{ x\mu - \sum_{k=1}^{x} \hat{\tau}_k \right\} \tag{2}$$

where

$$\gamma = \sum_{k=0}^{x} \exp\left\{ k\mu - \sum_{j=1}^{k} \tau_j \right\}$$

is a normalizing factor ensuring that

$$\sum_{x=0}^{m} p_x = 1$$

and where $\tau_x = 1, 2, \ldots, m$ are again m thresholds on the continuum. This model has been called the Rasch rating model because it has all the distinguishing properties of the Rasch model for dichotomously scored responses (see *Rasch Measurement Theory*). That is, the structural parameters of thresholds (and items) can be separated from the distributions of the groups (and persons). Figure 2 shows the response probability curves for three ordered categories. As with Equation (1), the exponent of Equation (2) can be modified to suit the particular data collection format. Thus for contingency tables, the exponent may take the form

$$x\mu_g - \sum_{k=1}^{} \tau_k$$

Alternatively, if the thresholds are considered differently spaced from group to group, it may be modified to

$$x\mu_g - \sum_{k=1}^{} \tau_{kg}$$

It is important to note one similarity and three differences between Equations (1) and (2). First, they both take the logistic form. Second, in Equation (1) the logit is given by the

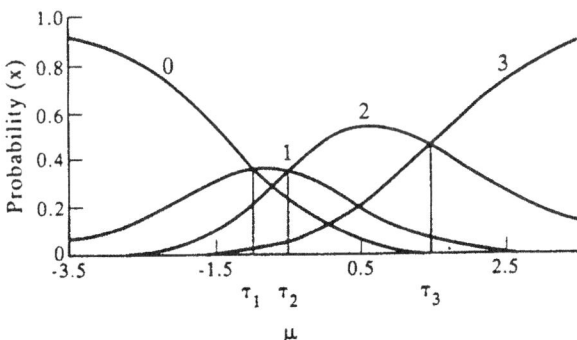

Figure 2   Probability of the response process with the Rasch cumulative threshold model.

ratio of cumulative probabilities while in Equation (2) it is given by the ratio of probabilities of *adjacent* categories: $(p_x/p_{x-1}) = \exp\{\mu - \tau_x\}$ from which $\ln(p_n/p_{x-1}) = \mu - \tau_x$. Third, the exponent in Equation (2) has parameters that are additive, there being no general discrimination α as in the exponent of Equation (1). Fourth, while the same term "threshold" is used in both models, because they are defined differently, they have different values.

When the MLE is used, the *sufficient statistics* for the estimate of μ_g is

$$r_g = \sum_{x=1} x f_{gx}$$

and the sufficient statistic for the estimate of τ_x is

$$t_x = \sum_{x=1} f_{gx}$$

Further, the solution equations for μ_g and τ_x respectively are

$$r_g = \frac{1}{\gamma} N_g \sum_x \exp\left\{ x\hat{\mu}_g - \sum_{k=1}^{x} \hat{\tau}_k \right\}, \, g = 1, \, G$$

$$t_x = \frac{1}{\gamma} \sum_g \exp\left\{ x\hat{\mu}_g - \sum_{k=1}^{x} \hat{\tau}_k \right\}, \, x = 1, \, m$$

These equations with the imposed constraints

$$\sum_g \hat{\mu}_g = 0 \text{ and } \sum_x \hat{\tau} = 0$$

must be solved iteratively because they are implicit, and not explicit, equations in the parameters.

The existence of sufficient statistics shows that they contain all the information about the parameters that is available in the responses. In addition, these statistics are the simple sums of the integers, which are exactly the ones used in the elementary measurement analogy. Through Equation (2), the constrained qualitative responses are mapped onto an additive or linear scale, which also estimates the thresholds. The simple total score r_g is then not seen as the sum of equally spaced thresholds as in the full measurement analogy, but as a count of the number of thresholds that have been passed. The weighting of the categories is taken account of by the threshold estimates. Andrich (1979) discussed the application of this model to contingency tables.

In the case of assessment of individuals, the location μ in the exponent of Equation (2) is again modified to include a person parameter β_n and a difficulty or affective value parameter δ_i so that $\mu_{ni} = \beta_n - \delta_i$ giving the logit

$$(\beta_n - \delta_i) - \sum_{k=1}^{x} \tau_k$$

Then the sufficient statistic for the person parameter β_n is

$$r_n = \sum_i x_{ni}$$

for the task or statement parameter δ_i it is

$$s_i = \sum_n x_{ni}$$

and for the threshold parameter τ_x it is

$$\tau_x = \sum_n \sum_i I_{nix}$$

where $I_{nix} = 1$ if the response is in category x, and 0 otherwise. That is, t_x is simply the total number of responses in category x across all tasks or statements and across all persons.

Again, the statistic r_n for estimating the person parameter is identical to that used in the elementary measurement analogy and shown by Likert (1932) to be satisfactory for the measurement of persons. In fact, because $\hat{\beta}_n$ and r_n are monotonically related, their correlation is nearly perfect. The transformation of r_n to $\hat{\beta}_n$ again maps or transforms the qualitative responses onto a linear scale.

If it is assumed that the threshold spacings, whatever they are, are not equal across tasks or statements, then the exponent of Equation (2) may be qualified further to

$$x\beta_n - \sum_{k=1}^{x} \tau_{ki}$$

(Wright and Masters, 1982). The total score

$$r_n = \sum_i x_{ni}$$

remains the sufficient statistic for the person parameter β_n, thus indicating the appropriateness of scoring successive categories with successive integers while taking account of variations in spacing between the thresholds (see *Partial Credit Model*).

5   Connecting ratings to measurements

Because the Rasch rating model does not require that the distances between thresholds are equal, the integer score on each rating is not itself a measure. However, the parameter estimates are on a linear scale and are measures up to an interval level. With respect to a conformable set of statements or tasks, the only difference between usual measures and those estimated through the Rasch model for ratings is one of degree. The precision can be increased by increasing the number of tasks or statements. Thus the Rasch rating model formally completes the measurement analogy.

A related aspect of the Rasch rating model is that it formalizes the popular and simple, yet theoretically weak, Likert approach to the attitude measurement of individuals. Likert had originally proposed his approach to circumvent the time-consuming requirement of scaling statements required by the approaches of Thurstone. The formalization permits the Likert approach to subscribe to all the rigorous requirements of Thurstone for scales (see Thurstone, 1927; Scale P), including the scaling of tasks or statements. In particular, and with respect to a conformable set of tasks or statements, any subset will lead to the same measure of a person. Similarly, the scale values of the statements and thresholds will be invariant across the abilities or attitudes of the persons measured. The linearity and this form of invariance are key aspects of Thurstone Scales. Andrich (1982) presented a full discussion of the way in which the Rasch rating model unifies the Thurstone and Likert approaches to scaling and measurement.

6   Quality control

It is stressed that these invariance properties hold in data only if they conform to the Rasch model; whether or not they do so conform, is an empirical question.

There are two related advantages in using the above explicit measurement model for analyzing ratings. The obvious one is that the measures – not simply ratings – become available. Second, and equally important, is that in the very process of attempting to obtain formal measurements, a greater understanding of the variable or trait in question should follow. A close examination of response patterns that do not conform to the rating model may be as informative in understanding the variable as when they do conform, and the Rasch rating model permits a refined analysis that detects lack of conformity in various ways.

6.1   The response pattern

One important and necessary feature for measurement, which can be checked, is whether the response pattern is internally consistent. According to the rating model, every person is expected to score higher on an easier question in achievement testing, or a smaller affective value in attitude measurement. If the ratings do not conform satisfactorily to this pattern, then a single measure to represent the responses cannot be justified. This perspective leads naturally to the analysis of profiles.

6.2   The "halo" effect

There can be many sources of inconsistency, as in all measurement data. When judges rate performances, a judge may gain an overall impression that affects all his or her ratings of the performance, even if these ratings are supposed to reflect different criteria. This is called the "halo" effect. The ratings in this case are "too consistent" as a result of the artificial dependence among them.

6.3   Rater leniency

In performance ratings, some raters may be more lenient or more harsh than others. If more than one rater rates with respect to a single task, then the rating model accounts for this effect. In particular, when δ is made to characterize the rater, it represents rater harshness. Alternatively, if more than one rater and more than one task are involved, then the logit in Equation (2) may be further modified to include the rater effect. For example, it may take the form

$$x(\beta_v - \delta_i - \eta_j) - \sum_{k=1} \tau_k$$

where η_j is the harshness of rater j. Other qualifications are also possible.

6.4   Response sets

Another possible systematic source of inconsistency occurs when different individuals use the categories differently. For example, some raters may use the extreme categories, while others may use the central categories, relatively too often. Both types of response patterns, reflecting what are known as response sets, which can threaten valid measurement, can be identified by the relative location of the thresholds.

6.5   Number of categories and the neutral category

In constructing rating response formats that will minimize the above problems, two further issues need to be considered. First, the number of categories should be large enough to take advantage of the judge's capacity to discriminate, but not greater. Guildford (1954) gave empirical evidence to guide the choice of the number of categories. Usually four or five are used for unipolar scales and five to nine for bipolar scales. Second, in bipolar scales, the Neutral or Undecided category has been the subject of much study. It seems not to attract responses consistent with those found on either side of it, it being a "catch-all" category in which people who do not understand the question, as well as people who are genuinely undecided, or neutral, respond. It seems best to construct statements that would attract few responses in this category, and then to exclude the category when the statements are used to obtain measures. Linacre (1989) covered these issues in greater detail.

7   Distinctions between the Rasch and Thurstone models

While the models of Rasch and Thurstone can be used for the same purpose, there are some major conceptual differences between the two, and in a well-defined sense they are incompatible specifications. First, as indicated already, the Rasch model permits a separation of the person and item parameters in the estimation, and therefore no assumptions about the distribution of the person parameters is necessary in any estimation. Such a separation is not possible in the Thurstone model. Second, because the thresholds are determined by the cumulative probabilities in the Thurstone model, the ordering of the estimates of the thresholds must be in the *a priori* order designated by the investigator. Therefore, any problem in the operation of the rating scale on this feature cannot be identified by the model, and all data, no matter how poorly collected, will always show the *a priori* ordering. In contrast, the threshold estimates in the Rasch model may be disordered relative to the intended ordering of the categories, indicating that something has gone wrong in the specification or operation of the categories. The reversal of the threshold estimates does not in itself indicate exactly what has gone wrong.

However, one example seems consistent: when a *neutral* category is inserted in a bipolar scale, the Rasch model consistently shows reversed threshold estimates, confirming other well-established evidence that this category does not operate on the same dimension as the other categories. Third, and most profound, is that in the Thurstone model it is possible to collapse adjacent categories after the data are collected, and the estimates of the thresholds between other categories will not be affected. In contrast, in the Rasch model, if the data collected with a particular set of categories conform to the model, then if a pair of adjacent categories are collapsed, the data will no longer conform to the Rasch model with one fewer category. The implication of this result is that the specification of the categories, and the collection of data in these categories, is an integral part of the construction of the variable, and that if the separation of the person and item parameters is to be retained, these cannot be tampered with after the data are collected. The full implications of this result have not been explored (see also: *Rasch Measurement Models*).

References

Andersen, E. B. 1977. Sufficient statistics and latent trait models. *Psychometri.* **42**(1), 69–81.

Andrich, D. 1978. A rating formulation for ordered response categories. *Psychometri.* **43**(4), 561–73.

Andrich, D. 1979. A model for contingency tables having an ordered response classification. *Biometrics* **35**(2), 403–15.

Andrich, D. 1982. Using latent trait measurement models to analyse attitudinal data: A synthesis of viewpoints. In: Spearrit, D. (ed.) 1982. *The Improvement of Measurement in Education and Psychology*. ACER, Melbourne. pp. 89–126.

Bock, R. D. 1975. *Multivariate Statistical Methods in Behavioral Research*. McGraw-Hill, New York.

Dawes, R. M. 1972. *Fundamentals of Attitude Measurement*. Wiley, New York.

Guildford, J. P. 1954. *Psychometric Methods*, 2nd edn. McGraw-Hill, New York.

Likert, R. 1932. A technique for the measurement of attitudes. *Archives of Psychology* **140**, 52.

Linacre, J. M. 1989. *Many-faceted Rasch Measurement*. MESA Press, Chicago, Illinois.

Samejima, F. 1969. Estimation of latent ability using a response pattern of graded scores. *Psychometric Monographs*. Vol. 34, No. 17, Part 2. University of Chicago Press, Chicago, Illinois.

Thurstone, L. L. 1927. Psychological analysis. *American Journal of Psychology* **38**, 368–89.

Wright, B. D. and Masters, G. N. 1982. *Rating Scale Analysis*. MESA Press, Chicago, Illinois.

9 Sufficient Statistics in Educational Measurement

E. B. Andersen

In social science (including educational) research, a sufficient statistic is a function of a given set of observations which extracts the available information about a given parameter. Under certain regularity conditions the statistical model must belong to a so-called exponential family to allow for a nontrivial sufficient statistic. In educational measurement sufficient statistics are found in the form of the raw score for a set of binary items, but also other scoring formulas are sufficient statistics. This article surveys these formulas.

1 Sufficient statistics

A sufficient statistic for an unknown parameter θ was defined by Fisher (1922) as a function $t(x_1, \ldots, x_n)$ for which the conditional distribution of any other statistic $t^*(x_1, \ldots, x_n)$ given $t(x_1, \ldots, x_n) = t$ is independent of θ. This is the most direct way to express that t extracts all the available information in the sample $(x_1, \ldots, x_n)$ about θ. When the value of t is known no other statistic t^* can provide further information about θ.

An equivalent definition is that the conditional distribution of $x_1, \ldots, x_n$ given the value of $t(x_1, \ldots, x_n)$ does not depend on θ. This can also be expressed as

$$f(x_1, \ldots, x_n | \theta) = h(x_1, \ldots, x_n | t) g(t | \theta) \tag{1}$$

where f is the joint density or probability of $x_1, \ldots, x_n$, h is the conditional density or probability given $t(x_1, \ldots, x_n) = t$ and g is marginal density or probability of t. The converse of Equation (1) is known as Neyman's criterion: If $f(x_1, \ldots, x_n | \theta)$ can be written as in Equation (1), where h does not depend on θ and g only depends on the x's through t, then t is sufficient for θ.

A statistic t_1 is called minimal sufficient if for any other sufficient statistic t, t_1 is a function of t.

2   Distributions admitting sufficient statistics

In the mid-1930s it was shown by Darmois (1935), Pitman (1936), and Koopman (1936) that under certain regularity conditions only so-called exponential families admitted nontrivial sufficient statistics. These regularity conditions were primarily of two types:

(a) $f(x_i|\theta)$ is twice differentiable in x_i

(b) The range of x_i, that is, those x_i for which $f(x_i|\theta)>0$, does not depend on θ.

Under (a) and (b) it can be shown, that the density $f(x_i|\theta)$ has the form

$$f(x_i|\theta) = c(\theta)\, \exp[v(x_i)\phi(\theta)]h(x_i) \qquad (2)$$

and that the minimal sufficient statistic is

$$t = \sum_{i=1}^{n} v(x_i) \qquad (3)$$

$\tau = \phi(\theta)$ is called the canonical parameter. For further references the reader should consult Brown (1964).

When the xs are discrete random variables, assumption (a) of course does not hold and must be replaced by other assumptions, essentially to ensure that t represents a true data reduction. This is a critical assumption as the observations themselves are sufficient. Papers by Andersen (1970) and Denny (1972) give more details.

But also for the discrete case, where $f(x_i|\theta)$ is the probability of getting the observation x_i, the expression in Equation (2) defines an exponential family, where t, given by Equation (3), is a minimal sufficient statistic.

In the case of the vector valued parameter $(\theta_1, \ldots, \theta_n)$ an exponential family takes the form

$$f(x_i|\theta_1, \ldots, \theta_k) = c(\theta_1, \ldots, \theta_k)\, \exp\left[\sum_{j=1}^{m} v_j(x_i)\phi_j(\theta_1, \ldots, \theta_k)\right]h(x_i) \qquad (4)$$

with m sufficient statistics

$$t_j = \sum_{i=1}^{n} v_j(x_i), \quad j = 1, \ldots, m \qquad (5)$$

for the canonical parameters $\phi_1, \ldots, \phi_m$. Note that m can be and often is less than k, which means that only m parameters can be estimated in a minimal sufficient way.

3   Sufficient statistics for latent trait models

In educational research, binary responses to n items, that is, $x_i = 0$ or 1, are considered, and the latent trait $p_i(\theta)$ of item i is defined by

$$f(x_i | \theta) = \begin{cases} p_i(\theta) & \text{for } x_i = 1 \\ 1 - p_i(\theta) & \text{for } x_i = 0 \end{cases}$$

The latent trait is thus the probability of $x_i = 1$ for latent parameter θ. The latent parameter is often an individual parameter, for example the ability of a given individual. When θ goes from $-\infty$ to $+\infty$ it is usually assumed that $p_i(\theta)$ goes from 0 to 1. It is not assumed that the x_is are identically distributed, but rather that $p_i(\theta)$ depends on certain item parameters (α_i, β_i). In Andersen (1977) it was proved that if there exists a minimal sufficient statistic $t(x_1, \dots, x_n)$ for θ, which does not depend on the item parameters, then $p_i(\theta)$ has the form

$$p_i(\theta) = \exp(\theta - \alpha_i)/[1 + \exp(\theta - \alpha_i)] \tag{6}$$

or what is known as the Rasch model. In this case the minimal sufficient statistic is the so-called raw score

$$t = \sum_{i=1}^{n} x_i \tag{7}$$

If t is allowed to depend on the item parameters, no results are available on the general form of $p_i(\theta)$, but the two-parameter logistic model

$$p_i(\theta) = \exp[(\theta - \alpha_i)\beta_i]/\{1 + \exp[(\theta - \alpha_i)\beta_i]\}$$

introduced by Birnbaum (Lord and Novick, 1968, Chapter 18) has the property that the weighted score

$$t = \sum_{i=1}^{n} \beta_i x_i \tag{8}$$

is sufficient for θ.

The concept of sufficiency is essential when working with latent trait models. When the statistic t is sufficient for θ, the inference concerning θ can be based on the frequency distribution of t in the population without losing information about θ. It is also essential that t represents a real data reduction as is the case for the raw score, Equation (7). For the weighted score, Equation (8), with odd parameter values $\beta_1, \dots, \beta_n$, there may only exist one response pattern $x_1, \dots, x_n$ for a specific value of t and the scoring does not represent a data reduction.

4   The polychotomous case

Andersen (1977) also discusses the case, where x_i is a categorical variable with $m > 2$ possible values. For this case, it can be shown that the score vector $(t_1, \ldots, t_m)$, where t_j = number of responses in answer category J, is a sufficient statistic, but not necessarily minimal sufficient. If θ is a one-dimensional latent parameter, it is thus of importance to find a real-valued statistic t to estimate θ with. As the vector $(t_1, \ldots, t_m)$ is sufficient, t will be a function of $(t_1, \ldots, t_m)$. Under certain assumptions on the kind of data reduction provided by t, it was shown by Andersen (1977) that t must have the form

$$t = \delta_1 t_1 + \ldots + \delta_m t_m$$

where $\delta_1 - \delta_2 = \delta_2 - \delta_3 = \ldots = \delta_{m-1} - \delta_m$. Such δ's are known as equidistant scorings of the response categories.

References

Andersen, E. B. 1970. Sufficiency and exponential families for discrete sample spaces. *J. Am. Statist. Assoc.* **65**, 1248–55.

Andersen, E. B. 1977. Sufficient statistics and latent trait models. *Psychbometri.* **42**, 69–81.

Brown, L. 1964. Sufficient statistics in the case of independent random variables. *Ann. Math. Statist.* **35**, 1456–74.

Darmois, G. 1935. Sur les lois de probabilité à estimation exhaustive. *C.R. Acad. Sci. Paris* **200**, 1265–66.

Denny, J. L. 1972. Sufficient statistics and discrete exponential families. *Ann. Math. Statist.* **43**, 1320–22.

Fisher, R. A. 1922. On the mathematical foundations of theoretical statistics. *Phil. Trans. Royal Soc. A* **222**, 309–68.

Koopman, B. O. 1936. On distributions admitting a sufficient statistic. *Trans. Amer. Math. Soc.* **39**, 399–409.

Lord, F. M. and Novick, M. R. 1968. *Statistical Theories of Mental Test Scores.* Addison-Wesley, Reading, Massachusetts.

Pitman, E. J. G. 1936. Sufficient statistics and intrinsic accuracy. *Proc. Camb. Phil. Soc.* **32**, 567–79.

Part II

Applications of Measurement in Research and Assessment

10 Adaptive Testing

D. J. Weiss and J. L. Schleisman

An adaptive test is an educational or psychological test in which the questions/items to be administered to an examinee are selected based on the examinee's responses to previously administered items (Wainer, 1990; Weiss, 1985; Weiss and Betz, 1973). This approach to test administration contrasts with a conventional test (e.g., a typical paper-and-pencil test) in which all examinees receive the same fixed set of items. Adaptive testing has also been called tailored (Lord, 1971), response-contingent, programmed, computerized, automated, individualized, and branched testing. Adaptive testing can be differentiated from sequential testing. In a sequential test (Kingsbury and Weiss, 1983), test items are administered either in a fixed order or are randomly selected from an item bank; in an adaptive test, items are selected from an item bank based on a prespecified item selection rule. Sequential and adaptive tests are similar in that the number of items administered to an examinee can vary, although they are based on different termination rules.

A variety of approaches has been developed for administering adaptive tests. In some approaches, a new item is selected after a single item is administered and scored. In others, a new block of items – sometimes called a "testlet" (Wainer and Kiely, 1987) – is administered based on an examinee's performance on a previously administered set of items. An examinee's response to an item (or set of items) is scored and the next item(s) administered is based on the examinee's response to the previous item(s). If the response is correct, the next item administered will be a more difficult item; if the response is incorrect, the next item will be less difficult. Items are selected from an item bank that contains items with predetermined difficulties (and perhaps discriminations and other item characteristics).

An individual taking an adaptive test will receive a set of items that is most appropriate for his or her performance level within the limitations of the item bank – the items selected are adapted to the characteristics of the examinee during testing. Adaptive tests are designed to improve measurement over conventional tests by improving the efficiency of test administration – by administering the minimum number of items necessary to measure each examinee – and by controlling the precision of measurement. Some adaptive tests have been designed for individual administration by a trained psychometrist (e.g., the

Stanford–Binet) while others have been proposed for paper-and-pencil administration, for example, Lord's (1971) flexilevel test. Computerized adaptive testing (CAT), in which test items are administered by an interactive computer, takes full advantage of the capabilities of adaptive testing.

In this entry the basic principles of adaptive testing are summarized and adaptive item selection procedures based on prestructured or unstructured item banks are described. In self-adapted testing, the examinee selects the difficulty of the next item based on feedback received from performance on previous items. Adaptive testing has been shown to increase measurement efficiency and precision. Adaptive testing has also been applied to personality assessment. Potential problems with adaptive testing and future research areas are also discussed.

1   Principles of adaptive testing

An adaptive test requires: (a) a precalibrated item bank, (b) a procedure for initial item selection, (c) a procedure for item selection during test administration, (d) a method for scoring the test, and (e) a procedure for terminating the test.

1.1   Initial and adaptive item selection

If the first item is allowed to vary for each examinee, the test has a variable entry point; if all examinees begin testing with the same item, the test has a fixed entry. Prior information from a teacher, the examinee, demographic data, or previous test data can be used to determine the initial item selection for a variable entry test. Any information that is known to be correlated with the performance or trait estimates derived from the adaptive test will reduce test length or improve performance or trait estimation.

The item selection rule differs depending on the type of adaptive test being used. Adaptive item selection procedures are based either on prestructured item banks, which typically implement fixed-branching procedures, or unstructured item banks which implement variable-branching procedures. Scoring methods and termination rules are generally implemented in the context of the prestructured and unstructured item bank dichotomy.

1.2   Prestructured item selection/fixed branching

Early approaches to adaptive testing were based primarily on fixed-branching procedures, such as the individually administered Stanford–Binet intelligence test (Terman and Merrill, 1960), originally developed by Alfred Binet in the early 1900s. Based on the examinee's responses to items at a given "mental age", items at another "mental age" are selected and administered to the examinee. Other types of adaptive tests that use prestructured item selection rules include two-stage (Lord, 1971), pyramidal (Hansen, 1969), flexilevel (Lord, 1971), and stratified adaptive (stradaptive) tests (Weiss, 1973).

Wainer and Kiely (1987) proposed a "multistage fixed-branching CAT model that substitutes multi-item 'testlets' for single items as the unit for test development and administration . . . . A testlet is a group of items related to a single content area that is developed as a unit and contains a fixed number of predetermined paths that an examinee may follow" (p. 190). Testlets (see also Kingsbury and Zara, 1989; Thissen *et al.*, 1989) can use a linear or hierarchical branching scheme. In a linear testlet, all examinees receive all items; in a hierarchical testlet, branching is based on previous responses. Testlets are combined to form a larger test. Similar to a two-stage test, an examinee who performed well on a previous testlet would be administered a testlet of higher difficulty; an examinee who responded poorly on items in the previous testlet would be administered a testlet of lower difficulty.

Although two-stage tests have minimal adaptive capability, Adema (1990) proposed mixed-integer linear programming models for constructing two-stage tests. Adaptive tests based on "testlets" are an implementation of the concepts underlying two-stage tests and are an extension of tests based on blocks of items with more than two stages and therefore have many of the limitations of the two-stage tests, even though they are implemented with more sophisticated models.

The fixed-branching item selection procedures have several limitations: (a) they do not use all information available from the items (i.e., they generally only make use of item difficulty and ignore other item characteristics such as discriminations and susceptibility to guessing); (b) they are based on arbitrary scoring methods; (c) with the exception of the stradaptive tests, they are based on a fixed termination criterion (i.e., all examinees receive the same number of items); and (d) the amount of adaptation for each examinee is limited.

1.3   Unstructured item banks/IRT-based adaptive testing

Variable-branching strategies based on item response theory (IRT), latent trait test theory, and item characteristic curve theory which operate from unstructured but precalibrated item banks, have been developed and represent the state of the art in adaptive testing. IRT is a family of mathematical models that describe the probability of a correct response as a function of item characteristics and examinee characteristics. The three-parameter IRT model (Hambleton and Swaminathan, 1985; Lord, 1980) characterizes items by the three parameters: difficulty, discrimination and pseudo-guessing; and examinees by using their performance or trait level. The Rasch (1960) or one-parameter (logistic) model describes items using only difficulty. The three item parameters can be combined into a single index called the "item information function", which describes how precisely an item measures at different performance or trait levels.

Item information is a primary criterion for selecting items in IRT-based adaptive testing. Two item selection procedures are commonly used: maximum information (Weiss and Kingsbury, 1984) and Bayesian (Owen, 1975). In maximum information adaptive testing, items are selected that provide maximum levels of item information at the examinee's currently estimated performance/trait level. The item is administered, and the examinee's performance/trait estimate is based on the responses to all previously answered test items using maximum likelihood estimation. The new performance or trait estimate is then used to select the next item to be administered to that examinee. This process is repeated until a termination criterion is reached.

Bayesian-based item selection uses Bayes's theorem to select the one item from all unadministered items that will minimize the Bayesian posterior variance of the performance/trait estimate after it is answered. Performance or trait level is then re-estimated using Bayesian estimation procedures, and the item bank is again searched to identify the unadministered item that minimizes the posterior variance. The procedure is repeated until some predetermined level of the Bayesian posterior variance is reached. This procedure permits more explicit use of prior information to determine starting points than does maximum information item selection. However, the use of prior information introduces biases into the scoring procedure that reduce levels of measurement precision for individuals whose performance or trait levels are not near the prior estimate (e.g., Weiss and McBride, 1984).

IRT-based adaptive testing addresses the limitations of and offers advantages over non-IRT-based adaptive testing. The item's difficulty, discrimination, and guessing parameters can all be taken into account; scoring methods are not arbitrary; and item banks can be used efficiently because there is no predetermined branching structure.

A final advantage of IRT-based adaptive tests is that the termination criterion can be tailored to the purpose of testing. IRT scoring procedures allow for both performance or trait estimation and the estimation of individualized errors of measurement. Consequently, in situations where it is important to test all examinees to a given level of precision, the termination criterion can be based on precision of measurement; given an adequate item bank, testing can be continued for each examinee until the desired level of precision is reached. In other testing applications (e.g., selection of individuals against some predetermined criterion level of performance, such as in job or educational admissions testing or mastery testing), it is necessary to test an individual only long enough to determine whether she or he is above or below the cutoff value (e.g., Weiss and Kingsbury, 1984).

The implementation of IRT-based adaptive testing involves substantial amounts of numerical calculation after each item is administered. Typically, calculations involve estimating the performance or trait level and its standard error, and the amount of information provided by each item at the current performance or trait estimate. Because of these computational requirements, IRT-based adaptive testing must be computer-administered and is the basis for most implementations of CAT.

1.4   Self-adapted testing

Self-adapted testing (SAT) (Kingsbury and Zara, 1989; Rocklin and O'Donnell, 1987) has been proposed as a method to reduce test anxiety, which, it is believed, will maximize an examinee's performance. This is accomplished by allowing the examinee to choose the difficulty level of the items administered. For example, after answering an item, an examinee would be informed immediately whether the answer was correct or incorrect. The examinee would then choose the next item. The items in SAT are prestructured by difficulty into levels or strata, therefore, it is very similar to the stradaptive testing procedure, except that the examinee selects the difficulty of the next item. An obvious problem with this procedure is that an examinee may select only items that she or he can answer correctly (i.e., the examinee may never select an item from a stratum containing items of higher difficulty). Although this might result in reduced test anxiety, it might not yield much psychometric information about the examinee.

2   Improved measurement efficiency and precision with adaptive testing

Research comparing conventional tests and adaptive tests has demonstrated that a major advantage of adaptive tests over conventional tests is increased measurement efficiency. Improved measurement efficiency occurs when test length is reduced, but measurements of comparable or superior quality are obtained. For example, McBride and Martin (1983), compared computer-administered conventional tests and CATs and found that for any given test length the adaptive tests had higher reliabilities than the conventional tests. A 10-item CAT had a slightly higher reliability than a 25-item conventional test (0.87 vs. 0.86). McBride and Martin also determined how many items a conventional test would need to reach a reliability equal to that of a CAT. For a 5-item adaptive test, the conventional test would need to be 2.57 times as long or 12.85 items in length to attain the same level of reliability as the CAT.

Adaptive tests not only provide increases in measurement efficiency, but also improved measurement precision. Adaptive tests provide measurements of equal or greater precision than conventional tests of the same length (Wise and Plake, 1989). McBride and Martin (1983) found that the alternate forms reliability of a 9-item adaptive test (0.800) was equal to that of a 17-item conventional test (0.798). Others (Olsen *et al.*, 1989; Weiss, 1982; Weiss and Vale, 1987; Wise and Plake, 1989) reported similar findings concerning measurement efficiency and precision. The decreases in test length of adaptive tests do not negatively affect validity; McBride and Martin found that scores on a 10-item adaptive test correlated 0.80 with a 50-item criterion test, but it took 30 items on a conventional test to reach the same level of validity.

3   Potential problems with CAT

Perhaps the most commonly cited potential problem with CAT is context effects (Bunderson *et al.*, 1989; Kingsbury and Zara, 1989; Wainer and Kiely, 1987). The effects of prior items in a test on succeeding items are known as "context effects". In conventional testing, each examinee receives the same items in the same order and, therefore, any context effects would exist for every examinee and should not differentially impact any particular examinee. In CAT, each examinee potentially receives different items and/or items in different orders. "Cross-information" (information that one item may inadvertently contribute about the answer to another) is a context effect. If Item 3 provided information that would help an examinee answer Item 5, then every examinee would have access to that information in a conventional test. However, in an adaptive test, only an examinee who received both Items 3 and 5 consecutively would have access to that information.

Another context effect is "unbalanced content", which refers to the situation in which there is repeated emphasis on a particular content area. For example, in an adaptive mathematics test designed to test the four basic mathematics skills (addition, subtraction, multiplication and division), an examinee may only be administered items which require subtraction and addition. This unbalanced sampling of the different content areas can occur if item selection is based solely on psychometric criteria, particularly in nonhomogeneous content domains.

Another potential problem with adaptive testing, termed "lack of robustness", results from the fact that adaptive tests are often shorter than conventional tests and thus lack redundancy (Bunderson et al., 1989; Wainer and Kiely, 1987). An item that is functioning poorly in an adaptive test might have a greater impact on test results than a poorly functioning item would have in a longer conventional test. This problem may be mitigated, however, by the fact that IRT-based CATs typically use item banks that consist primarily of items of high psychometric quality and are carefully analyzed before the tests are administered.

In a conventional test, items are usually presented in order of increasing difficulty. This procedure may give many examinees confidence by allowing them to experience success at the beginning of a test. In an adaptive test, the initial item administered is usually one of average difficulty. For a high-ability examinee, the items would increase in difficulty from the initial item administered; examinees of lower ability would probably experience failure on the initial items because they would be too difficult, and the subsequent items administered would decrease rather than increase in difficulty. This might cause frustration, anxiety, and lower performance in the lower ability group. This problem can be mitigated by the use of informed entry points based on appropriate prior information. Also, in a good adaptive test, all examinees will receive a set of items on which they will experience 50 per cent success; therefore adaptive tests may have the beneficial effect of equalizing the psychological test-taking environment for both high and low ability examinees.

Another potential problem is the effect that reviewing items and changing responses might have on performance or trait estimates. Wise and Plake (1989) concluded that unless denying examinees the opportunity to review items and to change responses "adversely affects the test scores of a meaningful proportion of examinees . . . they should probably not be provided" (p. 7). Lunz *et al.* (1992) studied the effect of allowing examinees to review items and to alter responses. They examined what the ability estimates (and standard errors) of those who were allowed to review would have been if they had not been allowed to review. They found that the ability estimates correlated 0.98, which suggests that being able to review did not significantly alter the ability estimates of examinees and that the standard errors of measurement were not affected by review.

4   Application of adaptive testing to personality assessment

Although the majority of adaptive testing research has occurred with ability and achievement measurement, the principles and procedures apply equally well to the measurement of personality variables (Butcher, 1987). For example, Waller and Reise (1989) compared two IRT-based adaptive testing strategies to a paper-and-pencil version of a personality scale. Their procedure involved "real-data simulation", namely, post-hoc simulation of adaptive tests using responses of examinees to a previously administered conventional personality scale. They found that the adaptive procedure resulted in a reduction in test length of as much as 50 per cent without any loss in measurement precision. Using a decision-based adaptive testing strategy, they obtained perfect success rate (i.e., no incorrect classifications) even though only 25 per cent of the items, on average, were administered.

Roper *et al.* (1991) administered paper-and-pencil and CAT versions of the MMPI-2 to college students. They were interested in determining whether score comparability was compromised because fewer items were administered in the CAT version. No significant differences were found across modalities even though fewer items were administered in the CAT version.

5   Future directions of adaptive testing

Adaptive testing, particularly CAT, moved from the research to the implementation stage during the early 1990s. Flexible microcomputer software for adaptive testing – the MicroCAT Testing System (Assessment Systems Corporation, 1988) which is capable of administering almost any type of adaptive test – has been developed and upgraded. Several major testing programs in the United States have converted large testing programs to CAT administration, including the Educational Testing Service's Graduate Record Examination and Scholastic Aptitude Test. The National Council of State Boards of Nursing began nationwide CAT administration of its nursing licensing examinations in 1994 and the public school system in Portland, Oregon has converted many of its achievement tests to CAT administration (Kingsbury, 1990).

Further research is needed to resolve the aforementioned potential problems described. New algorithms must also be developed and evaluated to integrate psychometric item selection criteria with content and other considerations, and research is needed on entry points and termination criteria for different testing applications.

Almost all CAT developments are based on power tests, yet speed is an important component of ability. There is a need to develop CAT models that will allow adaptation of examinee response latency or time of response presentation. In addition, almost all adaptive testing research has been based on multiple-choice items. However, both computer presentation of tests and the trend toward performance evaluation allow for a much richer and varied kind of test item than is possible with paper-and-pencil tests (e.g., Bunderson *et al.*, 1989). These new kinds of test items will pose new challenges for the development of CAT (see also: *Item Response Theory; Rasch Measurement Theory; Computerized Educational Testing; Rasch Measurement Models; Item Banking*).

References

Adema, J. J. 1990. The construction of customized two-stage tests. *J. Educ. Meas.* **27**(3), 241–39.

Assessment Systems Corporation. 1988. *MicroCAT 3.0* (Computer program). Assessment Systems Corporation, St. Paul, Minnesota.

Bunderson, C. V., Inouye, D. K. and Olsen, J. B. 1989. The four generations of computerized educational measurement. In: Linn, R. L. (ed.) 1989. *Educational Measurement*, 3rd edn. Macmillan, New York.

Butcher, J. N. 1987. *Computerized Psychological Assessment: A Practitioner's Guide*. Basic Books, New York.

Hambleton, R. K. and Swaminathan, H. 1985. *Item Response Theory: Principles and Applications*. Kluwer-Nijhoff, Boston, Massachusetts.

Hansen, D. N. 1969. An investigation of computer-based science testing. In: Atkinson, R. C. and Wilson, H. A. (eds.) 1969. *Computer-assisted Instruction: A Book of Readings*. Academic Press, New York.

Kingsbury, G. G. 1990. Adapting adaptive testing: Using the MicroCAT testing system in a local school district. *Educational Measurement* **9**(2), 3–6, 29.

Kingsbury, G. G. and Weiss, D. J. 1983. A comparison of IRT-based adaptive mastery testing and a sequential mastery testing procedure. In: Weiss, D. J. (ed.) 1983. *New Horizons in Testing: Latent Trait Test Theory and Computerized Adaptive Testing*. Academic Press, New York.

Kingsbury, G. G. and Zara, A. R. 1989. Procedures for selecting items for computerized adaptive tests. *Applied Measurement in Education* **2**(4), 359–75.

Lord, F. M. 1971. The self-scoring flexilevel test. *J. Educ. Meas.* **8**(3), 147–51.

Lord, F. M. 1980. *Applications of item Response Theory to Practical Testing Problems*, Erlbaum, Hillsdale, New Jersey.

Lunz, M. E., Bergstrom, B. A. and Wright, B. D. 1992. The effect of review on student ability and test efficiency for computerized adaptive tests. *Appl. Psychol. Meas* **16**(1), 33–40.

McBride, J. R. and Martin, J. T. 1983. Reliability and validity of adaptive ability tests in a military setting. In: Weiss, D. J. (ed.) 1983. *New Horizons in Testing: Latent Trait Test Theory and Computerized Adaptive Testing*. Academic Press, New York.

Olsen, J. B., Maynes, D. D., Slawson, D. and Ho, K. 1989. Comparisons of paper-administered, computer-administered and computerized adaptive achievement tests. *Journal of Educational Computing Research* **5**(3), 311–26.

Owen, R. J. 1975. A Bayesian sequential procedure for quantal response in the context of adaptive mental testing. *Journal of the American Statistical Association* **70**, 351–56.

Rasch, G. 1960. *Probabilistic Models for Some Intelligence and Attainment Tests*. Danmarks Paedagogiske Institut, Copenhagen.

Rocklin, T. and O'Donnell, A. M. 1987. Self-adapted testing: A performance-improving variant of computerized adaptive testing. *J. Educ. Psychol.* **79**(3), 315–19.

Roper, B. L., Ben-Porath, Y. S. and Butcher, J. N. 1991. Comparability of computerized adaptive and conventional testing with the MMPI-2. *Journal of Personality Assessment* **57**(2), 278–90.

Terman, L. M. and Merrill, M. A. 1960. *Stanford–Binet Intelligence Scale*. Houghton Mifflin, Boston, Massachusetts.

Thissen, D., Steinberg, L. and Mooney, J. A. 1989. Trace lines for testlets: A use of multiple-categorical response models. *J. Educ. Meas.* **26**(3), 247–60.

Wainer, H. 1990. *Computerized Adaptive Testing: A Primer.* Erlbaum, Hillsdale, New Jersey.

Wainer, H. and Kiely, G. L. 1987. Item clusters and computerized adaptive testing: A case for testlets. *J. Educ. Meas.* **24**(3), 185–201.

Waller, N. G. and Reise, S. P. 1989. Computerized adaptive personality assessment: An illustration with the absorption scale. *J. Pers. Soc. Psychol.* **57**(6), 1051–58.

Weiss, D. J. 1973. *The Stratified Adaptive Computertized Ability Test.* Research Report No. 73-3. University of Minnesota, Minneapolis Department of Psychology, Minnesota.

Weiss, D. J. 1982. Improving measurement quality and efficiency with adaptive testing. *Appl. Psychol. Meas.* **6**(4), 473–92.

Weiss, D. J. 1985. Adaptive testing by computer. *J. Consult. Clin. Psychol.* **53**(6), 774–89.

Weiss, D. J. and Betz, N. E. 1973. *Ability Measurement: Conventional or Adaptive?* Research Report No. 73-1. University of Minnesota, Department of Psychology, Psychametric Methods Program, Minneapolis, Minnesota.

Weiss, D. J. and Kingsbury, G. G. 1984. Application of computerized adaptive testing to educational problems. *J. Educ. Meas.* **21**(4), 361–75.

Weiss, D. J. and McBride, J. R. 1984. Bias and information of Bayesian adaptive testing. *Appl. Psychol. Meas* **8**(3), 273–85.

Weiss, D. J. and Vale, C. D. 1987. Adaptive testing. *Appl. Psychol.* **36**(3/4), 249–62 (Special Issue: Computerized Psychological Testing).

Wise, S. L. and Plake, B. S. 1989. Research on the effects of administering tests via computers. *Educational Measurement* **8**(3), 5–10.

11 Computerized Educational Testing

W. J. van der Linden

Test developers have always had a keen interest in automating their operations. The reasons for this interest are twofold. First, inherent in the use of objective tests is the attempt to standardize all steps in the testing process. Examples of such steps are processing the answer sheets, scoring the test, sending out reports, and filing the test scores. Even before computers were available, developers in charge of large-scale testing programs had well-defined routines to deal with these activities. Where possible, technical devices were used to reduce the amount of labor involved in running the programs – as was illustrated by the widespread use of the so-called Hollerith machine for processing punch cards until the 1960s. As soon as computers became available, most of the routines were immediately transferred to application software. Efficiency was further enhanced by combining the use of computers with optical scanners for reading answer sheets.

Second, educational testing relies heavily on psychometric analyses for the evaluation of the quality of its measurements. These analyses are necessary to provide information on the difficulty of the test items, the factorial structure of the test, the accuracy of test-based predictions, or to estimate the transformation that equates scores on new test forms to the scale of an old form. The analyses must usually be done under time pressure – after the test is administered, but before it can be scored. In the pre-electronic era, ingenious computational schemes and mechanical calculators were employed to make efficient use of the amount of time available. Small wonder, then, that test specialists were among the first to use the computer for running their analyses.

The applications of the computer to educational testing outlined above will not be reviewed in this entry. Though convenient, they are straightforward and represent no major shift in technology. Rather, the interest will be in the *new* testing technology generated by the introduction of the computer in the field of educational testing. This technology capitalizes not just on the computer's ability to process data at high speed, but also on the interactivity it offers, the immediate access to mass storage of data, the novel use of graphics and the power to control multimedia environments. Examples of this new technology can be found in such applications as adaptive or response-contingent testing, innovative use of graphics to implement new response test formats, measurement of response time, computerized test design, authoring systems for test-item writing, use of test scores in support systems for instructional decision-making, and the simulation of real-life processes and the use of multimedia environments to increase test validity. This new technology is rapidly expanding. It has influenced the work of test developers, profoundly, and has already reached the schools through software for personal computers (PCs).

It is important to note that all these applications are not the sheer result of the availability of powerful computer hardware or the introduction of efficient programming methodology. A fortunate coincidence was the simultaneous development of a completely new class of models in test theory. The theory underlying these models is known as Item Response Theory (IRT). IRT supports the use of computers in educational testing in a natural way. It can be used to provide each item stored in the computer with independent estimates of their properties, to have the computer select tests from item banks, to govern item administration procedures, or to design models for scoring tests with new computer-based response formats. The last example shows that IRT has not only made computerized testing feasible, but that the use of computers in testing has also stimulated the development of psychometric models that would neve have existed without the presence of computers. The principles on which IRT models are built and their role in computerized testing will be briefly outlined below.

The remainder of the entry has been organized into five sections, each dealing with a different aspect of computerized testing: (a) computerized adaptive testing; (b) item banking systems; (c) automated test design; (d) authoring systems for item writing; (e) multimedia testing.

Due to developments in information technology, computer hardware quickly becomes obsolete. Therefore, no attempt will be made to review available hardware. For the same reason, existing computer software will be referred to only incidentally. Reviews of hardware and software, however, can be found in Baker (1989), Hsu and Yu (1989), and Roid (1989).

1   Computerized adaptive testing

Since the first standardized test was developed, test designers have been aware that they must solve the following paradox. Suppose a test has to be designed to measure the ability of an examinee in a certain domain. If the ability of the examinee were known, it would be possible to design a perfect test. None of the items the examinee would pass or fail for sure would be included in the test; it would be more efficient to just fill out the answer sheet for these items in advance. Instead, the test would concentrate on those items for which it would be uncertain if the examinee could pass. A test with this content would not waste any time, but be maximally informative about the ability of the examinee. However, the ability of the examinee is *unknown*; otherwise there would be no reason whatsoever to administer a test.

It is known that Stanford and Binet tried to solve this paradox for their first intelligence test, published in 1905, by continuously adapting the content of the test to the ability level of the examinee during the session. A description of the adaptive aspect of the Stanford–Binet test is given in Reckase (1989). Others have followed the same path, introducing such procedures as two-stage testing and flexi-level testing.

In two-stage testing, the format of a single test is replaced by a short routing test along with a series of second-stage tests of varying difficulty. The routing test is administered first. It is quickly scored, and then the scores are used to select an optimal second-stage test. Of course, the scores on the routing test have restricted reliability. In fact, the above dilemma now takes the form of a trade-off between the lengths of the routing test and the second-stage tests. But it can be shown that scores based on the combination of a routing test and an optimal second-stage test generally are more reliable than scores on a single test of the same total length. A more ingenious design offers the flexi-level test. With a flexi-level test the examinee codes his or her answer by scratching a certain area on the answer sheet, and then finds information on the next item to continue with.

Both the two-stage and flexi-level testing procedure are described in Lord (1980). Though these procedures are more efficient than traditional standardized tests, they suffer from such constraints as a restricted number of items in the test form and a fixed order in which the difficulties have to be present in the test.

For testing to be fully adaptive, the following conditions have to be realized:

(a) a delivery system that is able to retrieve items quickly from a large pool and display them to the examinee;
(b) a rule for scoring the ability of the examinee on the basis of the responses to the items already displayed;
(c) a rule for selecting the next item to be displayed as a function of the score on the previous item;
(d) a rule for selecting the item in the pool to start the testing procedure with; and
(e) a rule for stopping the procedure.

A computer, with its capacity for storing huge amounts of text and graphics, its quick access to memory, its video monitor for displaying the items, its keyboard for recording the responses, and its enormous power to perform calculations according to complex rules, forms an ideal environment for adaptive testing. Hence it was not until mass production of computers started that adaptive testing became practicable.

Nowadays most test publishers sell adaptive versions of their standardized tests. Adaptive tests are delivered in the form of software for PCs along with a data file that contains the test items. A typical file has 200–500 items. It is not uncommon to find that computerized adaptive testing (CAT) realizes the same accuracy with 30–40 per cent of the items needed previously in roughly 50 per cent of the time. The gain in time can be used to test abilities in a larger series of domains and increase the predictive validity of the test. A favorite application in education is therefore testing for placement in educational programs; for example, at college entry level.

Introductions to CAT can be found in Bunderson *et al.* (1989), Lord (1970, 1980), van der Linden and Zwarts (1989), Wainer *et al.* (1990) and Weiss (1983).

1.1   *IRT models*

As was observed above, to implement a CAT procedure, in addition to a computer environment, rules are needed to select the items, score the examinees, and stop the procedure. At the same time, these rules need statistical information about the items on which to operate. It is here that IRT plays its role.

The power of IRT resides in the fact that it parameterizes the ability of the examinees and the properties of the items *separately*. Due to the presence of these separate parameters, estimates of the item parameters can be used to correct for the properties of the items when scoring the ability of an examinee. This correction is an instance of statistical adjustment. It allows scoring of different selections of items from a pool on the *same* scale – a feature on which CAT capitalizes by adapting the properties of the items to a current estimate of the examinee's ability.

A graphical representation of an IRT model is given by a response curve, as in Figure 1. The curve displays the probability of a correct response on item i as a function of the value of ability parameter of the examinee. The item parameters determine the shape and the location of the curve. A common mathematical form of the curve is given by the so-called three-parameter logistic formula:

$$P(+|\theta) = c_i + (1 - c_i)[1 + \exp\{-a_i(\theta - b_i)\}]^{-1} \tag{1}$$

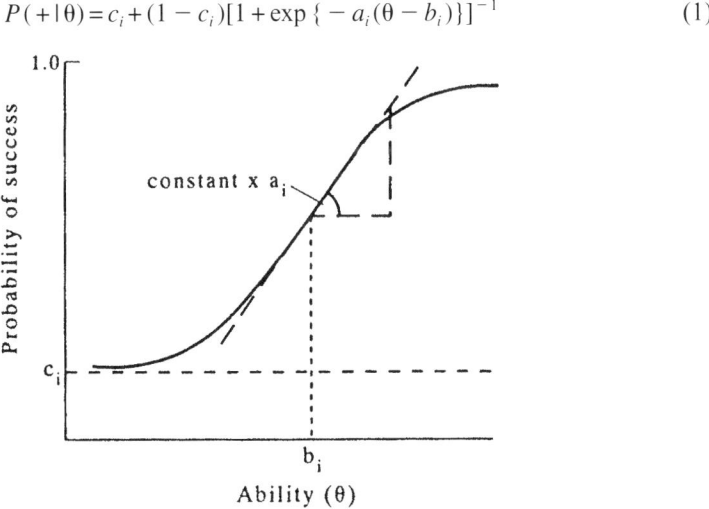

Figure 1   Response curve of a test item.

where θ is the ability of the examinee, b_i is the difficulty (location), a_i the discriminating power (slope), and c_i the guessing parameter (lower asymptote) of item 1.

A practical feature of IRT models is the presence of a measure generally known as the information function. An information function gives the information in the responses to an item or a test as a function of the true ability of the examinee. For five items as well as the test composed of the items, the information functions associated with the model in Equation (1) are given in Figure 2. Conceptually, it is convenient to view information functions as a related but more fundamental concept than classical reliability. Unlike classical reliability, information functions are no single coefficients but functions showing how "reliably" the test or the item measures examinees at each possible value of the ability scale. It should be observed that the information functions for the five items in Figure 2 sum to the function for the test. Also, the reciprocal of the square root of the information function is a measure that gives the standard error of measurement as a function of the ability value.

Full introductions to IRT models are given in Hambleton and Swaminathan (1984), Lord (1980), and Rasch (1980) (see *Item Response Theory*).

1.2   Rules for CAT

An obvious way to select items in a CAT procedure is to base the selection of subsequent items on the information functions of the items in the pool. Each next item could then be selected such that it has maximum information at the ability value where the examinee is estimated to be. In fact, this *maximum information* rule is the most popular item

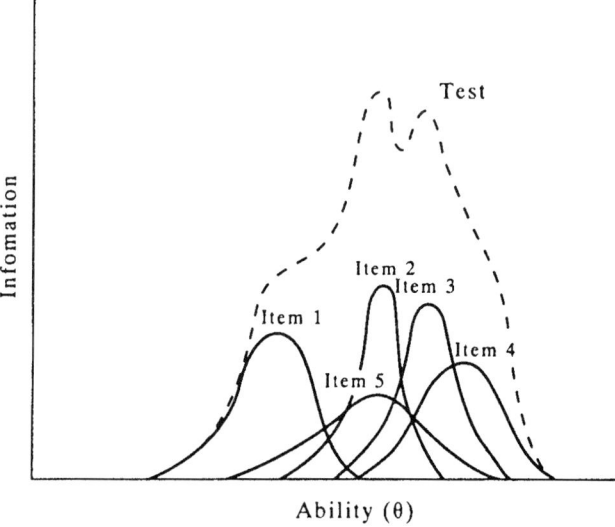

Figure 2   Item and test information functions.

assignment rule in CAT. It is not the only rule; an alternative is a Bayesian rule in which the next item is selected such that the expected variance of the posterior distribution for the examinee's ability value is minimal (see Owen, 1975).

The two item assignment rules can only be operational if an estimate of the examinee's ability from his or her responses to previous items is obtained in realistic time. A natural partner of the maximum information rule is *maximum-likelihood estimation* of ability. In this estimation method, the ability value that maximizes the likelihood of the response pattern obtained from the examinee is defined to be his or her ability estimate. The use of modern hardware and software gives quick estimates for the IRT models currently in use. For Owen's rule, Bayesian ability estimation is the appropriate choice.

The choice of the first item in a CAT procedure is important, because considerable gain of information, and hence reduction of test length, can be obtained if the first item is not too far off target. If no prior information about the ability of the examinee is present, the best choice is to start with an item that is optimal at a (subjective) estimate of the location of the ability distribution of the population of examinees the test has been designed for. If prior information is available, for example, in the form of information on background variables with a known regression on the ability variable, better choices are possible. It is also possible to have the computer collect information on the background variables and update the regression function with each new examinee.

A CAT procedure is stopped if enough information on the examinee's ability is gathered. Figure 3 gives an account of the information gathering process for maximum-likelihood estimation of ability. For convenience, confidence intervals derived from the test information function are depicted, rather than the information functions themselves. For each item added to the test, the width of the interval decreases. At the same time the location of the interval converges to a certain ability value. As soon as the width of the interval is below a prespecified threshold value, the procedure stops.

Presenting the above examples of CAT rules, it has been assumed that all item parameters in the IRT model have been estimated earlier. This point will now be taken up.

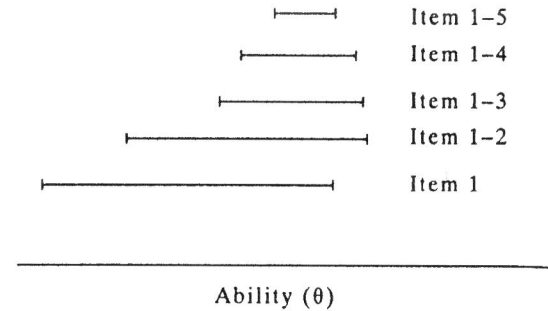

Item 1–5
Item 1–4
Item 1–3
Item 1–2
Item 1

Ability (θ)

Figure 3   Increase of accuracy of ability estimation in CAT.

2   Item banking systems

The traditional instrument in educational measurement is the standardized test. Standardized testing is based on the paradigm of experimental control as it has evolved in the literature on experimental design earlier in the twentiech century (van der Linden, 1986b). The steps involved in developing a standardized test are as follows. First, a blueprint of the test is established that will guarantee a test with appropriate content validity. Then the items are written and field-tested. If the final version of the test is assembled, another field test is done to crossvalidate the reliability of the test and to estimate norm distributions. After this, the test is fixed and any change of content will invalidate its use.

Though of practical value, standardized tests are inflexible and their use is often less than optimal. For example, the same test can never be used twice for the same group of examinees without running into security problems. Also, if a test has to serve a population of examinees varying widely in ability, it is likely to be suboptimal for most examinees. This is the paradox of test design referred to earlier. A related point is the problem of test use in educational research with a pretest–posttest design. If the treatment in this design is effective, a standardized test is likely not to be – one test can never reliably measure large differences in ability. More fundamentally, standardized tests may have low content validity, as the number of items in the test is usually much smaller than the number of items that defines the domain to be tested.

A new conception of educational measurement is test item banking. The paradigm of test item banking is not experimental control, but the idea of statistical adjustment already alluded to earlier (van der Linden, 1986a). In test item banking a computer is used to store a pool of items covering the content domain. In addition to the item, initial estimates of their parameters, like those defined by the IRT model in Equation (1), are stored. If the item banking system is operational, it functions as in the diagram in Figure 4. First, items are retrieved from the system to assemble tests. If test delivery is on-line, CAT is an attractive choice. However, if on-line delivery is not possible, tests can also be assembled in the form of paper and pencil tests tailored to their specific applications. Though not as efficient as CAT, such customized tests are much better than standardized tests. The tests may be of any length not exceeding the size of the item bank. They may be selected to be very difficult or easy, or to cover a broad or a small range of ability. Whatever the selection

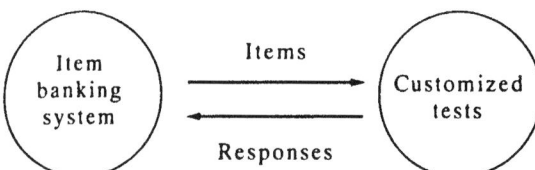

Figure 4   Diagram of test item banking system.

of items, when scoring the test, the item parameters in the system automatically correct the ability estimates of the examinees for the incidental properties of the items in the test.

Second, all responses on the items in the tests are fed back into the system. They are kept, for example, for a periodic update of the estimates of the item parameters. This feature allows the system to improve the fit of future tests with the users' specifications. Another possible use of item responses is to combine them with information on background variables to update estimates of norm distributions for various populations of examinees or to find optimal start items for use in CAT procedures.

Item banking systems cannot be run without the use of a computer. They need the use of database software, a text editor, and graphics software to store, catalog, and retrieve test items and their statistical information. If printed versions of tests have to be delivered, desktop publishing software is needed to edit the tests into their final form. Statistical software for parameter estimation and update, test scoring, and test assembly requires the power of modern processors. In fact, the combination of all these resources into a single application makes item banking systems unique. An item banking system that shares many of the features reviewed here is MicroCAT (Stone, 1989).

3 Automated test design

The process of designing customized tests assembled from an item banking system can be automated. As has already been demonstrated in Figure 2, information functions have the property of additivity. This feature allows the following design process.

First, the test constructor decides on a *target* for the information function of the test. The first step in reaching this decision is to determine what interval of ability values the test should measure. Then the shape of the target is chosen. The shape commonly depends on the intended use of the test. Figure 5 gives two possible targets, one for a diagnostic test, the other for a mastery test.

Second, items are selected from the pool to fill up the area below the target. This selection is subject to various constraints on the content of the items to guarantee a valid test.

To automate the design process, linear programming modeling can be used. A simple example shows the principles involved. The items in the pool are represented by index $i = 1, \ldots, N$. For each item a decision variable $x_i \in \{0, 1\}$ is defined to indicate whether or not the item will be selected for the test. The information in item i at ability value θ_k is denoted by $I_i(\theta_k)$. Now, let $T(\theta_k)$ be the value for the target information function of the diagnostic test in Figure 5 at ability values θ_k for $k = 1, \ldots, K$. Then, for an arbitrary test, non-negative numbers u_k and v_k can be used to denote the positive and negative slack between the actual information in the test at the values θ_k and the values of the target, $T(\theta_k)$. The following linear programming model selects the items from the pool such that the sum of the values of the slack variables is minimal:

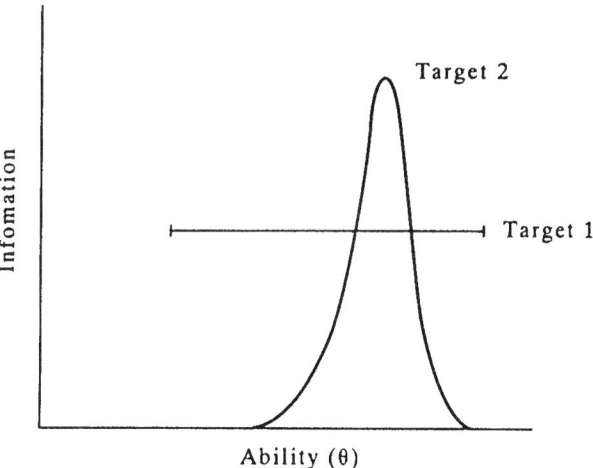

Figure 5   Two-target information functions in automated test design[a].

[a] 1 = diagnostic test, 2 = mastery test.

$$\text{minimize} \sum_{k=1}^{K} (u_k + v_k) \tag{2}$$

subject to

$$\sum_{i=1}^{N} I_i(\theta_k)x_i - u_k + v_k = T(\theta_k), \quad k = 1, \ldots, K \tag{3}$$

$$\sum_{i=1}^{N} x_i = 1 \tag{4}$$

$$u_k, v_k \geqslant 0, \quad k = 1, \ldots, K \tag{5}$$

$$u_i \in \{0, 1\}, \quad i = 1, \ldots, N \tag{6}$$

The objective function in Equation (2) minimizes the sum of values of the slack variables. In Equation (3) the slack variables are defined as the positive and negative deviation of the test information function from the target values. The only constraint added to the model is the requirement in Equation (4) which states that the length of the test be equal to n items. The possible values of all variables in the model are defined in Equations 5 and 6.

Linear programming software can be used to solve the model for the values of the variables that minimize the objective function and satisfy all the constraints. These values identify an optimal test.

In its present form, the above model is too simple to be realistic. In practice, additional constraints with respect to test content, cognitive level, such attributes as item readability or gender orientation, or various statistical criteria are needed to produce acceptable results. A catalog of constraints to deal with these and other specifications is given in van der Linden and Boekkooi-Timminga (1989). The paper also lists alternative objective functions. Theunissen (1985) presented a model that minimizes the length of the test. Another possibility is the simultaneous selection of a set of parallel tests (Adema, 1992; Boekkooi-Timminga, 1987). These references are only to a small portion of the methodology available now. It is recommended that the models always be applied in an interactive mode in which the computer program makes a proposal and the test constructor reviews the results, deciding which items to retain and which the computer has to replace. The process is iterated until a satisfactory result is obtained.

A software program that can be used to solve a model of minimal test length is *Optimal Test Design* (Verschoor, 1991). A general-purpose test assembly program with an interface that allows the user to make his or her own selection of objective function and constraints is *Contest* (ProGamma, 1993).

4   Authoring systems for test item writing

Item writing is a time-consuming process, mostly with several stages of review by content experts and field testing. It is surprising to note that the quality of the first versions of the items is often far less than optimal. It is not unusual to find that more than half of the items never reach the final stage. Typical errors involve poor use of language, response options with equivocal truth values, or the inability to produce items within a desired range of difficulty. As the quality of the items imposes an upper bound on the quality of the test, technology to improve the item writing process would be a major step forward.

Though the use of computers has penetrated the art of item writing, its potential has not been fully realized yet. Some of the developments will be briefly outlined here. In so doing, it is not the intention to suggest that the human item writer should be replaced by computer software. Human involvement will always be needed to produce text, make critical decisions, or approve products. However, computer technology may provide a convenient environment for item writers to fulfill all those functions.

A first level of computerized item writing involves the use of word processors to draft and edit test items. It was only recently that the use of word-processing programs added efficiency to the item writing process. The first reason for this fact is the amount of text in items usually is not large. Even a language test seldom has more than a few pages of text. Further, most test items have nontext materials, such as graphs, diagrams, drawings, mathematical symbols or expressions, or high-resolution photographs. Fortunately, nowadays most word processors have a spellcheck or thesaurus option and work with graphics programs for drawing graphs, preparing diagrams, or creating mathematical expressions. It is the integration of all these facilities that has made the word processor an attractive tool for item writing.

Though word processing may increase the efficiency of item writing, it is unlikely that the mere use of word processing will improve the effectiveness of the items produced. One possible exception is the factor that repeated use of the same word-processor may lead to better standardization of materials and hence to multiple test forms that better realize the ideal of parallel tests.

Significant improvement in the quality of item writing may be reached with a next generation of word-processing software, which will check for errors in grammar and style. Most item formats are sensitive to errors in the use of grammar and style. Examples of critical errors are: (a) inconsequent continuation of grammar in alternatives of multiple-choice items; (b) nonuniform style in checklist items, or between alternatives in matching items; and (c) failure to meet the requirement that fill-in items should allow wrong answers compatible with the grammar or style of the text presented.

Ultimate support of item writers would be provided by authoring systems specially designed for item writing. Such systems would integrate the above two levels of support with several new facilities to create an environment that fully supports each step in the item production process. The following facilities are attractive.

(a) The first is support of the selection of item format; for example, through a dialogue leading the item writer to an appropriate format for the application at hand.

(b) Second is guidance through the item writing process once a format is chosen. Options include the definition of a template for each individual item, lists of suggestions for typical text elements in items, routines for including nontext materials, procedures for ordering or randomizing alternatives or making layout decisions. Ideally, at each decision the system would provide the item writer with projections of the effects of the alternatives on the expected difficulty of the item.

(c) Third is the check of quality of the items. This includes not only the checks for spelling and grammar, but also for features such as the readability of the item and typical item writing errors that now figure in extensive checklists circulating in the testing literature on various item formats. A third level of item checking would be a search of collections of items for clues in one item to right answers on other items.

(d) Finally there is project management. Item writing typically is a process with many participants. If installed on networks of computers, authoring systems could provide communication between the participants and standardization of their output. In principle, the next participant could immediately continue working on a half product delivered by his or her predecessor.

A review of item writing software is given in Roid (1989), whereas Rikers (1988) described the design of an ideal authoring system for item writing.

5   Multimedia testing

For some content domains, test developers have had a tradition of using media other than paper and pencil to enhance the validity of their tests. Examples are the use of audio equipment in tests of listening comprehension in foreign languages and the use of slide presentations, for example, in traffic licensing exams.

With the availability of new electronic media, it is possible that these trends will expand. With built-in processors or in a computer-controlled mode, these media have the power to represent real-life situations convincingly in the test or to simulate processes in real time. Examples of these media are videodisc, CD-I, and special-purpose simulators. Their potential for use in educational testing has not been fully explored, but their impact on the validity of tests may be enormous. The first applications can be found in computer simulations of patient–management problems that are now introduced to test knowledge and skills in medical programs around the world. It is anticipated that other applications will soon follow. In particular, the use of computer-controlled multimedia environments in testing is attractive. Digital technology has reached the point where these environments are capable of simulating almost any process in "virtual reality". Applications of this technology are waiting to be used in educational testing. The only current obstacle is the cost of these media, which tend to prohibit large-scale application in education. However, if prices decrease, the number of applications in testing can be expected to increase immediately.

6   Conclusion

The use of information technology will no doubt further penetrate educational testing. Its effect on the flexibility and measurement quality of test procedures is enormous. At an increasing rate the field of educational testing will show tests delivered in novel formats and at unusual places. It is expected that this increased flexibility will provide a major impetus to further integration of educational testing and instruction (see also: *Adaptive Testing; Item Response Theory.*

References

Adema, J. J. 1992. Methods and models for the construction of weakly parallel tests. *Appl. Psychol. Meas.* **16**(1), 53–63.

Baker, F. B. 1989. Computer technology in test construction and processing. In: Linn, R. L. (ed.) 1989. *Educational Measurement*, 3rd edn. Macmillan, New York.

Boekkooi-Timminga, E. 1987. Simultaneous test construction by zero-one programming. *Methodika* **1**(2), 101–12.

Bunderson, C. V., Inouye, D. K. and Olsen, J. B. 1989. The four generations of computerized educational measurement. In: Linn, R. L. (ed.) 1989. *Educational Measurement*, 3rd edn. Macmillan, New York.

Hambleton, R. K. and Swaminathan, H. 1984. *Item Response Theory: Principles and Applications*. Kluwer-Nijhoff, Boston, Massachusetts.

Hsu, T.-C. and Yu, L. 1989. Using computers to analyze item response data. *Educational Measurement: Issues and Practice* **8**(3), 21–8.

Lord, F. M. 1970. Some test theory for tailored testing. In: Holtzman, W. H. (ed.) 1970. *Computer-assisted Instruction, Testing and Guidance*. Harper and Row, New York.

Lord, F. M. 1980. *Applications of Item Response Theory to Practical Testing Problems*. Erlbaum, Hillsdale, New Jersey.

Owen, R. J. 1975. A Bayesian sequential procedure for quantal response in the context of adaptive mental testing. *Journal of the American Statistical Association* **70**(350), 351–6.

ProGamma. 1993. *Contest* (Software and manual). ProGamma, Groningen.

Rasch, G. 1980. *Probabilistic Models for Some Intelligence and Attainment Tests*, rev. edn. University of Chicago Press, Chicago, Illinois.

Reckase, M. D. 1989. Adaptive testing: The evolution of a good idea. *Educational Measurement: Issues and Practice* **8**(3), 11–15.

Rikers, J. A. H. N. 1988. *Towards on Authoring System for Item Construction*. Department of Education, University of Twente, Entschede.

Roid, G. H. 1989. Item writing and item banking by microcomputer: An update. *Educational Measurement: Issues and Practice* **8**(3), 17–20.

Stone, C. A. 1989. Testing software review: MicroCAT. *Educational Measurement: Issues and Practice* **8**(3), 33–8.

Theunissen, T. J. J. M. 1985. Binary programming and test design. *Psychometri.* **50**(4), 411–20.

van der Linden, W. J. (ed.) 1986a. Special issue on test item banking. *Appl. Psychol. Meas.* **10**(4), pp. 325–443.

van der Linden, W. J. 1986b. The changing conception of measurement in education and psychology. *Appl. Psychol. Meas.* **10**(4), 325–32.

van der Linden, W. J. and Boekkooi-Timminga, E. 1989. A maximin model for test design with practical constraints. *Psychometri.* **54**(2), 237–47.

van der Linden, W. J. and Zwarts, M. A. 1989. Some procedures for computerized ability testing. *Int. J. Educ. Res.* **13**(2), 175–87.

Verschoor, A. 1991. *Optimal Test Design* (Software and manual). Cito, Arnhem.

Wainer, H. *et al.* (eds.) 1990. *Computerized Adaptive Testing: A Primer*. Erlbaum, Hillsdale, New Jersey.

Weiss, D. J. (ed.) 1983. *New Horizons in Testing: Latent Trait Test Theory and Computerized Adaptive Testing*. Academic Press, New York.

Further reading

Boekkooi-Timminga, E. and van der Linden, W. J. 1994. *Linear Models for Optimal Test*. Sage, Newbury Park, California.

Gutkin, T. B. and Wise, S. L. (eds.) 1990. *The Computer and the Decision-making Process*. Erlbaum, Hillsdale, New Jersey.

Hambleton, R. K., Zaal, J. N. and Pieters, J. P. M. 1991. Computerized adaptive testing: Theory, applications and standards. In: Hambleton, R. K. and Zaal, J. N. (eds.) 1991. *Advances in Educational and Psychological Testing*. Kluwer, Boston, Massachusetts.

12 Measurement of Developmental Levels

M. Wilson

For the purposes of this entry, a *developmental level* is a step in a sequence postulated as part of a theory of the progression of an individual toward maturity, broadly or narrowly defined. Most educational examples are in the field of cognitive development of persons, where the measurement tasks are designed to tap into the psychological structure of cognitive processes, but there are also theories involving developmental levels in the social and organizational fields. The measurement of developmental levels, then, includes the realization of such a sequence of levels in the real world and the techniques used to guide that realization. This entry is divided into three parts. There is a description of theories that involve developmental levels, followed by a description of methods of collecting data according to such theories, which is then followed by a discussion of various measurement models that may be used to understand and guide the measurement process. It should be noted that in standard form these models are distinguished from statistical models that may be used to analyze the resulting measurements. However, recent work in multilevel models has indicated ways to integrate measurement models and analysis models (Adams, Wilson and Wu, 1996). The entry concludes with a brief discussion of future directions in the measurement of developmental levels.

The changes in the individual toward maturity may be fundamentally continuous, in which case the levels may constitute no more than convenient labels for parts of the continuum, or the progression may be fundamentally discrete, in which case the levels must be conceptualized as a systemic part of the changes that occur. These two extremes are often termed the quantitative and the qualitative views respectively. Much debate has focused on this distinction especially in the Piagetian literature where the term "stage" was used to distinguish the qualitative extreme (see Brainerd, 1978, and the peer commentary that followed). In this entry, a view and a methodology are sought that span both the extremes, and the ground between.

1 Theories that involve developmental levels

The most prominent historical example of a theory that involves developmental levels is that of Piaget. This has led to the advance of a number of neoPiagetian theories such as those of Case and K. Fischer, and others outside of the cognitive area, such as the moral and social development theories of Kohlberg and Turiel. Although the Piagetian tradition has been the most prolific in its production of developmental levels, they have also appeared in other traditions, such as van Hiele levels in geometry learning (van Hiele, 1986), and Gagné's (1962) learning hierarchies. Biggs and Collis (1982) have devised the Structure of the Learning Outcome (SOLO) taxonomy to classify and structure hierarchically the responses of the learner to cognitive tasks ranging from descriptive essays to mathematics problems. According to the SOLO theory, the hierarchical levels of learning are that the response can (a) incorporate just one object of the stimulus, (b) incorporate several objects of the stimulus, (c) recognize the relations among these objects, and (d) go beyond the relationship implied in the stimulus. Clearly such a scheme can be applied in a wide variety of circumstances, and its range can be extended developmentally by "stacking" one taxonomy on another where the relations of one become the objects of the next one up.

Another related movement within education which is a rich source of developmental levels is the application of a cognitive science approach to students' conceptions of subject matter. This concentrates on discovering qualitative changes in students' conceptions of a phenomenon and thus very frequently results in the delineation of a sequence of developmental levels. In Europe, this is related to "phenomenography" (Marton, 1981), and is paralleled by a similar movement in the United States that has taken as one of its major techniques the study of conceptual differences between experts and novices. This latter tendency has meant that "relatively speaking, we have most knowledge of differences between beginners and experts, but less knowledge of the intermediate stages and the nature of the transitions from level to level" (Glaser *et al.*, 1987, p. 77). As the breadth of attention moves beyond the expert and novice extremes, more discussion can be expected of developmental levels in the American research, although different theoretical perspectives and labels may obscure their recognition (cf. White, 1988). More recent developments along these lines have occurred under the banner of "constructivism".

2   Measurement methods

The ultimate aim of measurement is to gain "objectivity and simplicity" (Glaser *et al.*, 1987, p. 64) in relating real world observations to theory. The concept of "developmental levels" may be a useful intellectual tool in this process in a range of different situations. When a developmental continuum is postulated and data are to be gathered, levels may provide a convenient way to label portions of the continuum for practical purposes, or may be used to give practical expression to a researcher's doubts about the precision of measurement (i.e., as an alternative to simply reporting standard error of measurement). At the other extreme, as in much qualitative investigation, researchers may be striving to preserve the full impact of individual differences in development by avoiding standardization of the measurement process. Nevertheless, at a conceptual level, the description of some sort of hierarchical structure becomes all but inevitable as the researcher proceeds to interpretation. This structure may be more complex than that of a sequence of developmental levels; often a structure like a (mathematical) lattice would seem more appropriate. Here a sequence of developmental levels may prove useful if considered as a projection of this more complicated structure on a single dimension. In between these two extremes lie the approaches that are explicitly based on developmental levels such as those described above.

In the search for simplicity, a sequence of developmental levels can assist by allowing the classification of individual differences into two importantly different types. First, there are those individuals whose behavior is consistent with that of the sequence of levels, and who can therefore be assigned to an estimated position in the sequence. Thus, an appropriate measurement approach must have techniques for identifying and displaying systematic ordering. Second, there are those whose behavior is inconsistent with the sequence, the interpretation of which requires theory beyond that establishing the sequence. Thus, an appropriate measurement approach must have, as a characteristic, techniques for identifying such individuals.

Researchers in the cognitive and phenomenographic traditions have used a range of qualitative assessment methods to guage an individual's position with respect to developmental levels. These include ratings of open-ended interviews, think-aloud protocols, observations of students' attempts to solve special tasks, and, in the late twentieth century, the results of students' interactions with computer programs. Piaget's *méthode clinique* is closely related to these.

Researchers working within the American tradition have tended more toward "objective" types of methods that closely resemble traditional types of tests. The most common has been labeled the "method of multiple tasks" by Fischer *et al.* (1984). Several tasks are developed for each level, each task is of a standardized format with a standardized method of judging success or failure, and all tasks are given to every individual for whom they are neither absurdly difficult nor absurdly easy. For assignment of individuals to levels, a decision rule, such as "highest level for which at least four tasks are judged correct out of five", must be provided. A particularly simple form of this method is the situation where there is just one task per level, in which case the decision rule takes on an equally simple form. Fischer *et al.* (1984) have described straightforward data summary methods to display the likely form of results from this method that would be consistent with discontinuities between levels.

A method that can be seen as a compromise between these two is the "partial credit" method. In this approach, a set of tasks is designed at which students may succeed at different levels, and a rating scheme is applied to identify the appropriate levels. It should be noted that the partial credit method shares some of the features of the phenomeno-graphic approach through the possibility of qualitative ratings, and some of the features of the method of multiple tasks through the possibility of replication and standardization of the ratings. One possible reason why it has not been more widely used is that models for polytomous data have historically been perceived as statistically rather complicated.

These methods can be used in a number of ways depending on the researcher's aims. For example, a researcher looking for a second-order discontinuity will have to use these methods with more than one variable. A very important distinction must be made between longitudinal designs and cross-sectional ones. Although the developmental focus of "developmental levels" obviously makes the longitudinal design the more desirable, practical considerations will work to keep cross-sectional designs common. The important issue is to be aware that the conseqeunces of these two designs for data analysis and interpretation are quite different. Fischer *et al.* (1984) give an example for the method of multiple tasks. The longitudinal design, since it is an extension of the cross-sectional design, can be used to address a wider variety of issues such as the characteristics of and influences on individual growth.

3   Measurement models

Although the results of applying the methods described above can be summarized using standard tabular and graphical devices, the interpretations commonly rest on a set of *ad hoc* assumptions to overcome the arbitrary features of the design. For example, in an application of the method of multiple tasks to the learning of a geometry sequence, Usiskin (1982) used five tasks per level, but then had to find a way of constructing a decision rule to classify the students. His solution was to experiment with two "at least *n* correct within each level" rules like the one described above. He chose "*n*" to be either three or four and conducted extensive studies of the differing effects of these two rules on both the internal consistency of the data he had collected and of the relationships of the resulting classifications with external validation measures. These exhaustive researches resulted in a global decision as to which one of these two rules was to be used. However, it left unanswered a great many other troubling questions, such as: why should the "*n*" be consistent from level to level, are all items equally representative of their levels, and, are all students who show inconsistency with the classification rule equally inconsistent? The analysis could not answer these questions, because the analysis itself was based on assumptions (i.e., that each item was an accurate indicator of level) that denied the possibilities raised in the questions.

One strategy that has been used to deflect attention from such disturbing measurement questions is to reduce the method of multiple tasks to its simplest form: use just one task per level. Here the decision rule seems incontestable: either the student is correct on that one task, and hence at least at the matching level, or the student fails the task and is somewhere below. The usual scientific principle of replication would also seem to be ignored by this strategy. Guttman's (1950) scalogram model provides a formal model of such data, although it was originally formulated for attitude scaling. This model has been used extensively in analyses of developmental levels data, but its usefulness has been criticized. Kofsky (1966) notes:

> It is also possible that the scalogram model may not be the most accurate picture of development, since it is based on the assumption that an individual can be placed on a continuum at a point that discriminates the exact skills he has mastered from those he has never been able to perform . . . A better way of describing individual growth sequences might employ probability statements about the likelihood of mastering one task once another has been or is in the process of being mastered (pp. 202–3).

She explains this need for a probabilistic approach thus: "The fundamental operations tested may indeed be mastered in a fixed sequence, yet the variations in instruction and material from task to task may have such a strong effect on performance that the regularities are masked" (Kofsy, 1966, p. 203). An extension of the idea of a scalogram that allows for some more flexibility in the direction of complicated learning hierarchies is ordering theory (Bart and Krus, 1973).

To answer these criticisms, it is necessary to go beyond the classification of responses into discrete ordinal classes and allow a probabilistic interpretation by mapping response patterns into levels. This provides some range in the difficultes of questions that represent particular levels, and allows for the possibility of transitional states when classification may be difficult (Biggs and Collis, 1982). To develop such an approach it is not necessary to do away with assumptions about the psychological and measurement processes underlying the tasks. However, other strong assumptions must be made that are based on a body of pre-existing validation evidence, and that potentially allow the checking of crucial assumptions in any application. This is the way to gain the objectivity mentioned by Glaser above.

One way to accomplish this is to base measurement on a latent trait approch such as the Rasch model (Rasch, 1960; Wright and Stone, 1979). The advantages of such an approach are: (a) the resulting metric provides a means of directly comparing the ability of the learner with the tasks representing the developmental levels; (b) the probabilistic formulation constitutes a framework within which response patterns can be interpreted beyond those that conform to the very restrictive rule such as the "at least n correct" rule discussed above; (c) there is no restriction in having a particular number of tasks at all levels; and (d) the probabilistic interpretation provides a scale of acceptability for response patterns, from those which are very consistent with the postulated sequence of levels to those which are very inconsistent with the sequence (see *Rasch Measurement Theory; Rasch Measurement Models*).

When the method of multiple tasks is used to assess a sequence of developmental levels with dichotomous items and the data are analyzed with the Rasch approach, the resulting scale would be expected to exhibit segmentation and order (Wilson, 1989b). Segmentation occurs when items representing different levels are contained in separate segments of the scale, with a nonzero distance between segments. Order occurs when the sequence is in the order predicted by the substantive theory. There are two ways that the analysis could indicate a problem. First, the item types could have been segmented, but in a different order from that predicted. This would imply that the theory itself, or perhaps its specifications for realization, are seriously flawed. The second and more common occurrence is that the two item types overlap. This can indicate that the item types do not accurately reflect the different parts of the learning sequence, or that some characteristic of the aberrant items, thought to be irrelevant, has influenced their difficulty in some way, or it can indicate a flaw in the theory. The Rasch scaling itself does not resolve which explanation is correct. The researcher must examine the context and theory for an explanation of the meaning of such discrepant results. More complex models which involve a discrimination parameter (e.g., Birnbaum and 3PL models) (see *Item Response Theory*) allow for different orderings of items at different abilities, which means that the concepts of segmentation and ordering cannot be defined. The Rasch approach also allows the investigation of person fit, which, under conditions of segmentation and order, is an assessment of which individuals are behaving consistently and inconsistently with the theoretical order. Thus, the two requirements for measuring developmental levels, as described above, can be examined in the context of Rasch models (see *Rasch Measurement Models*). Examples of the application of the Rasch model in Piagetian contexts have been provided by Bond (1996a), Dawson (1996) and Noelting *et al.* (1995). Bond (1996a, 1996b) and Bond and Bunting (1996) have also provided a comprehensive argument in favor of use of the Rasch model in such contexts.

This type of application of latent trait models has been criticized as lacking in an explicit role for the psychological structure underlying the developmental sequence (Spada and Kluwe, 1980). A model that allows the postulation of type and number of cognitive operations that are applied by members of certain manifest groups of persons, and includes these explicitly into the measurement model, is the Linear Logistic Test Model (LLTM) (Fischer, 1974, 1977; Spada, 1977). A summary of research and applications in this area is given by Spada and McGaw (1985). Another approach has been the differentiation of components of the various cognitive processes, strategies, and knowledge stores that are involved in item responses. Separate estimates of the parameters of these component processes can then be obtained and used in the Multicomponent Latent Trait Model (MTLM) (Embretson, 1985). Embretson (1985) has also described a model that combines these two. Applications of these models are usually restricted to the examination of just one pair of successive developmental levels, but the importance of their theoretical contributions should not be seen as diminished by that.

Another criticism that has been made of these models is that they do not allow second-order interactions among persons and tasks, which might, for example, result in different relative difficulties of items for different persons (Spada and McGaw, 1985). In one attempt to come to grips with this more difficult problem, Embretson (1985) described an extension of the MLTM that allows different components to be combined in varying ways through alternative solution strategies. A somewhat different approach was taken by

Wilson (1989a), who studied the patterns of change in relative item difficulties for persons at different levels of development, and explored the consequences for developmental theories. This approach is described in greater detail in the next section. A related approach was described by Rost (1991). Mislevy and Verhelst (1990) have placed this work in the broader context of the design of psychometric models that are suitable when subjects employ different solution strategies.

With the partial credit method of measurement, the developmental structure is built into the polytomous scoring scheme. In this case, the Partial Credit Model (Masters, 1982) is the member of the Rasch family that would be appropriate. Masters (1991) has described the application of this model to developmental levels in the context of a phenomeno-graphic example. Another polytomous model that can be used is the graded response model (Samejima, 1969) (see *Partial Credit Model*).

An interesting variant to this is where bundles of items are constructed, in which each bundle is composed of items logically connected to different developmental levels. An example is the "closed form" SOLO superitem (Biggs and Collis, 1982) where, for each superitem, a single piece of stimulus material is used to generate a sequence of tasks, one at each succssive SOLO level. There are several ways to model such a situation (Wilson, 1988). LLTM-type models pertaining to this context have been developed by Adams and Wilson (Adams and Wilson, 1992; Wilson and Adams, 1995). These have been applied to repeated trials in an intelligent tutoring system context by Pirolli and Wilson (1992, 1993).

The models described above have been based on a continuous latent trait conceptualiza-tion of the measurement model underlying the developmental levels. This is not the only such approach. One alternative is to consider each level as a class, leading to a latent class approach where the student is assumed to conform to exactly one of some small number of discrete latent classes. Although most applications of this idea have been applied to progress from one level to another (e.g., Macready and Dayton, 1980; Haertel, 1984), the possibility of applying this idea in more complicated situations has also been demonstrated (Rindskopf, 1983; Yamamoto, 1987).

Most of the models described above are principally concerned with cross-sectional methods of data collection. Certainly, once the measurement problem has been adequately addressed with cross-sectional data, it would seem that an appropriate next step, that would allow the examination of Fischer *et al.*'s (1984) second-order discontinuities, is the statistical analysis (using statistical growth models and/or multivariate analysis tech-niques) of the data produced by the measurement. An example using a hierarchical linear model is given in Bryk and Raudenbush (1987, p. 151). However, this approach may be an oversimplification of the problem. Fischer (1987) has argued strongly for an approach to the measurement of change based on the Rasch model using a relaxed version of the LLTM. Some researchers see the need to fold the measurement problem in with the longitudinal and multivariate structural model of the data (Jöreskog and Sörbom, 1979) and longitudinal factor analysis is an established area of research (Jöreskog and Sörbom,

1979; Meredith and Tisak, 1982; Tisak, 1984). The difficulties of applying factor analysis to data from developmental levels have been discussed by Fischer (1974), and Rogosa (1985) has strongly criticized this procedure in the context of structural equation modeling. Nevertheless, advocates of this approach argue for its usefulness (Short *et al.*, 1984), and several influential studies have been published (McCall *et al.*, 1977; Baltes *et al.*, 1978). Multidimensional scaling has also been applied to developmental levels data, and a survey of those applications is provided by Shoben and Ross (1987). Although these approaches tend not to emphasize the concept of developmental levels, the points debated are an important background for the measurement of change in developmental levels. Recent work integrating explicit measurement models of the type discussed in this entry and statistical models of growth (Adams, Wilson and Wu, 1996), may lead to some resolution of these methodological disputes.

4   Saltus: measuring across a discontinuity

Consider a situation where it is suspected that cognitive development occurs in spurts. In particular, suppose that there are two cognitive levels, A and B, where, according to the developmental theory, the level represented by A is cognitively precedent to that represented by B, and that this ordering is nontrivial. If the levels are operationalized by a certain number of items in each, then what would the data be expected to look like? The data would be expected to "chunk" together, with subjects succeeding on most of the A items and/or most of the B items.

One way to introduce a measure of discontinuity into a logistic model is to incorporate an interaction parameter τ_{hk} into the standard equation for a Rasch model. This is called the Saltus model (Wilson, 1989a):

$$\text{Probability }(y_{ij}=1\,|\,\varphi_{hi}=1) + \frac{\exp(\beta_i - \delta_j - \tau_{hk})}{1+\exp(\beta_i - \delta_j + \tau_{hk})}$$

where y_{ij} is the response (1 representing success, 0 representing failure) of person i in level h to item j in level k, and β_i and δ_j are the usual Rasch person and item parameters respectively and where $\varphi_{hi}=1$ indicates that person i is in level h. Note that the interaction is a group-level interaction between all persons at a certain level and all items at a certain level. Because of required identification constraints, there can be only one such parameter when there are two levels, which is labeled τ. The magnitude of τ, called the "gap" between person level B and item level A, indexes a discontinuity in the underlying latent trait. The larger τ, the greater the discontinuity: as τ approaches zero, the discontinuity vanishes and the Saltus model becomes a Rasch model. A possible representation of a hierarchical step is shown in Figure 1. The gap between A and B is evident. The asymmetry of growth is expressed by the dual location of the A level item: AA labels the response curve for persons at level A, AB for persons at level B. The solid lines are the modeled probabilities, the dashed lines indicate unused portions of the curves. It should

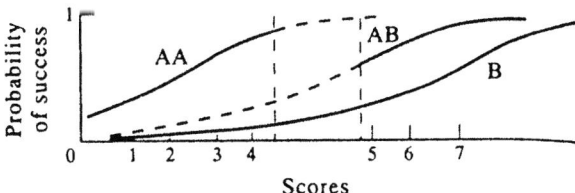

Figure 1   Saltus response curves for typical hierarchical items.

be noted that between scores 4 and 5, probabilities jump for the B items and that the relative success rates for the two item levels are not symmetrical as they would be for a Rasch model solution.

An example can be given using a re-analysis of data originally collected by Siegler (1981). He postulated a series of rules to describe development on balance beam tasks. Rule I is: Choose the side with greater weight; if weights are equal, choose neither. Rule II is: Same as I except that if weights are equal, choose the side with greater distance; if distances are also equal, choose neither. Rule III is the same as for II, but if neither the weights not distances are equal, muddle through. Rule IV is the same as using the correct formula. For the step from Level I to Level II, which Siegler represented by item sets D and S respectively, the Saltus results are illustrated in Figure 2. The following features of the results should be noted: (a) the large gap, as expected, between scores 4 and 5; (b) how success on S items increases considerably across the gap; and (c) how performance on D items deteriorates somewhat before returning to a high level of success. One might speculate that this last pattern might be a feature of the integration of an old skill into a new one. For the step from Levels II and III to Level IV, represented by the two item sets S and CE, the Saltus analysis is illustrated in Figure 3. Here the gap has become smaller, and, in fact, the Saltus estimates indicate that the gap is less than one standard error in size. A contributing factor in this was the wide range of difficulty for the CE items, indicating a heterogeneity in the item set that challenges the Siegler rule classification. The effect of

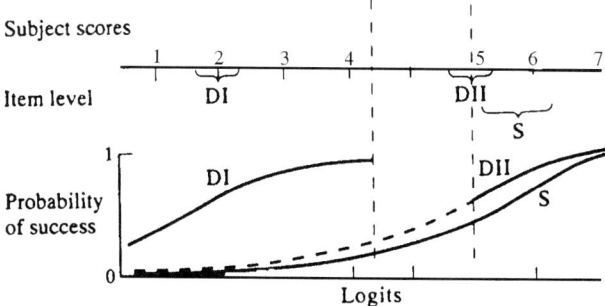

Figure 2   Saltus representation of the step from Siegler's rule I to rule II.

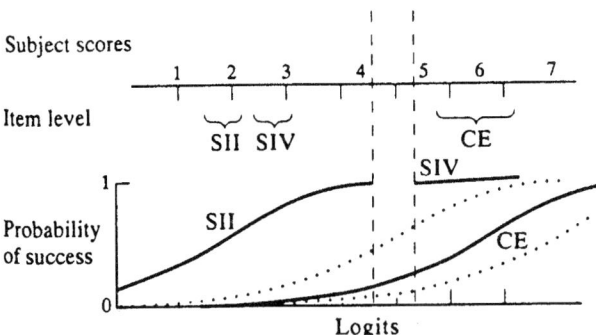

Figure 3   Saltus representation of the step from Siegler's rule II to rule IV.

this on the gap is illustrated by considering the easiest and hardest CE items, indicated in Figure 3 as dotted curves. For these extremes, the change in success rate across the gap increases from about 0.0 to about 0.7, an increase which indicates that there are differences among the CE items which have a strong influence on the empirical results. Wilson (1989a) should be consulted for more details of this example. Mislevy and Wilson (1996) have put this approach on a sound statistical basis, and an interesting example has been published by Demetriou *et al.* (1993). Applications involving polytomous student responses (which allow for correspondingly more complex interpretations) have been given by Wilson and Draney (1995) and Draney, Wilson and Pirolli (1996).

5   Future directions

The predominance of the cognitive science approach to psychology and education could transform the style of measurement that is used by the community of educational researchers and practitioners. Will the psychometric and educational measurement communities adapt to these new demands? This entry has attempted to describe the concept of "developmental level" as a common approach. For those applying cognitive science perspectives, the central idea of change in cognitive structure can lead quite naturally to developmental levels. For psychometricians, developmental levels offer a metaphor for interesting and complex models that articulate qualitative change on the basis of a quantitative substratum.

   This entry has described measurement models that are presently available and that resolve problems associated with earlier nonprobabilistic approaches used in the measurement of developmental levels. But more remains to be done. There has been little attention paid in the measurement literature to the problem of individual diagnosis which is so much a feature of the work of qualitative researchers (see, e.g., Molenaar and Hoytink, 1987; Wright and Stone, 1979, on person fit). An important development has been the scrutiny by substantive researchers of measurement (Glaser *et al.*, 1987). The effect of this may be seen in examples such as the close examination of the conditions under which spurts in development can be measured, by Fischer *et al.* (1984), and the incorporation of a cognitive science style format for measurement in the component tasks described by Embretson (1985) and Wang, Wilson and Adams (in press) (see also: *Rasch Measurement Theory*).

References

Adams, R. and Wilson, M. 1992. A random coefficients multinomial logit: Generalising Rasch models. Paper presented at the annual meeting of the American Educational Research Association, San Francisco, California.

Adams, R., Wilson, M. and Wu, M. 1996. Multilevel item response models: An approach to errors in variable regression. *Journal of Education and Behavioral Statistics.* **22**(1), 47–76.

Baltes, P., Nesselroade, J. and Cornelius, S. 1978. Multivariate antecendents of structural change in development: A simulation of environmental patterns. *Mult. Beh. Res.* **13**(2), 127–52.

Bart, W. and Krus, D. 1973. An ordering-theoretic method to determine hierarchies among items. *Ed. Psychol. Meas.* **33**(2), 291–300.

Biggs, J. and Collis, K. 1982. *Evaluating the Quality of Learning: The SOLO Taxonomy.* Academic Press, New York.

Bond, T. 1996a. Piaget and measurement I: The twain really do meet. *Archives de Psychologie* **63**, 71–87.

Bond, T. 1996b. Piaget and meawsurement II: Empirical validation of the Piagetian model. *Archives de Psychologie* **63**, 155–85.

Bond, T. and Bunting, E. 1996. Piaget and measurement III: Reassessing the méthode clinique. *Archives de Psychologie* **63**, 231–55.

Brainerd, C. 1978. The stage question in cognitive-developmental theory. *Behav. Brain Sci.* **1**(2), 173–82.

Bryk, R. and Raudenbush, S. 1987. Application of hierarchical linear models to assessing change. *Psych. Bull.* **101**(1), 147–58.

Dawson, T. 1996. Moral reasoning and evaluative reasoning about the good life: A Rasch analysis of Armon's 13-year life-span investigation. Paper presented at the 26th annual symposium of the Jean Piaget Society, Philadelphia.

Demetriou, A., Efklides, A., Papdaki, M., Papantoniou, G. and Economou, A. 1993. Structure and development of casual-experimental thought: From early adolescence to youth. *Dev. Psychol.* **29**, 480–97.

Draney, K., Wilson, M. and Pirolli, P. 1996. Using the RCML model to investigate linear logistic test models in a complex domain. In: Engelhard, G. and Wilson, M. (eds.) *Objective Measurement III: Theory into practice.* Ablex, Norwood, New Jersey.

Embretson, S. 1985. Multicomponent latent trait models for test design. In: Embretson, S. (ed.) *Test design: Developments in Psychology and Psychometrics.* Academic Press, Orlando, Florida.

Fischer, G. 1974. *Einführung in die Theorie psychologischer Tests.* Huber, Bern.

Fischer, G. 1977. Linear logistic test models: Theory and application. In: Spada, H. and Kempf, W. (eds.) *Structural Models of Thinking and Learning.* Huber, Bern.

Fischer, G. 1987. Applying the principles of specific objectivity and of generalisability to the measurement of change. *Psychometri.* **52**(4), 565–87.

Fischer, K., Pipp, S. and Bullock, D. 1984. Detecting discontinuities in development: Methods and measurement. In: Emde, R. and Harmon, R. (eds.) *Continuities and Discontinuities in Development.* Plenum, New York.

Gagné, R. 1962. The acquisition of knowledge. *Psychol. Rev.* **69**(4), 355–65.

Glaser, R., Lesgold, A. and Lajoie, S. 1987. Toward a cognitive theory for the measurement of achievement. In: Ronning, R., Glover, J., Conoley, J. and Witt, J. (eds.) *The Influence of Cognitive Psychology on Testing.* LEA, Hillsdale, New Jersey.

Guttman, L. 1950. The basis of scalogram analysis. In: Stouffer, S., Guttman, L., Suchman, F., Lazarsfeld, S., Star, S. and Clausen, J. (eds.) *Studies in Social Psychology in World War II. Vol. 4: Measurement and Prediction.* Princeton University Press, Princeton, New Jersey.

Haertel, E. 1984. Detection of a skill dichotomy using standardized achievement test items. *J. Educ. Meas.* **21**(1), 59–72.

Jöreskog, K. and Sörbom, D. 1979. *Advances in Factor Analysis and Structural Equation Models*. Abt, Cambridge, Massachusetts.

Kofsky, E. 1966. A scalogram study of classificatory development. *Child Dev.* **37**, 191–204.

Macready, G. and Dayton, C. 1980. The nature and use of state mastery models. *Appl. Psychol. Meas.* **4**(4), 493–516.

Marton, F. 1981. Phenomenography – describing conceptions of the world around us. *Ins. Sci.* **10**(2), 177–200.

Masters, G. 1982. A Rasch model for Partial Credit scoring. *Psychometri.* **47**, 149–74.

Masters, G. 1991. *The Measurement of Conceptual Understanding*. Australian Council for Educational Research, Hawthorn, Australia.

Meredith, W. and Tisak, J. 1982. Canonical analysis of longitudinal and repeated measures data with stationary weights. *Psychometri.* **47**(1), 47–67.

McCall, R., Eichorn, D. and Hogarty, P. 1977. Transitions in early mental development. *Mon. Soc. Res. Ch. Dev.* **42**(3), 108.

Mislevy, R. and Verhelst, N. 1990. Modelling item responses when different subjects employ different solution strategies. *Psychometri.* **55**(2), 195–215.

Mislevy, R. and Wilson, M. 1996. Marginal maximum likelihood estimation for a psychometric model of discontinuous development. *Psychometri.* **61**(11), 41–71.

Molenaar, I. and Hoytink, H. 1987. *The many null distributions of person fit indices*. European Meeting of the Psychometric Society, Enschede, The Netherlands.

Noelting, G., Coudé, G. and Rousseau, J. 1995. Rasch analysis applied to multi-domain tasks. Paper presented at the 25th annual symposium of the Jean Piaget Society, Berkeley, California.

Pirolli, P. and Wilson, M. 1992. Measuring learning strategies and understanding. A research framework. In: Frasson, C., Gauthier, G. and McCalla, G. I. (eds.) *Intelligent tutoring systems. Proceedings of the Second International Conference, Montreal.* Springer-Verlag, Berlin.

Pirolli, P. and Wilson, M. 1993. Knowledge and the simultaneous conjoint measurement of activity, agents and situations. Proceedings of the Cognitive Science Society Annual Meeting, Boulder, Colorado.

Rasch, G. 1960. *Probabilistic Models for some Intelligence and Attainment Tests*. University of Chicago Press, Chicago, Illinois.

Rindskopf, D. 1983. A general framework for using latent class analysis to test hierarchical and nonhierarchical learning models. *Psychometri.* **48**(1), 85–97.

Rogosa, D. 1985. Satisfying a simplex structure is simpler than it should be. *J. Ed. Stat.* **10**(2), 99–107.

Rost, J. 1991. A logistic mixture distribution for polychotomous item responses. *Br. J. Math. Stat. Psychol.* **44**(1), 75–92.

Samejima, F. 1969. Estimation of latent ability using a response pattern of graded scores. *Psychometric Monographs* No. 17, Vol. **34**(2), 100.

Shoben, E. and Ross, B. 1987. Structure and process in cognitive psychology using multidimensional and related techniques. In: Ronning, R., Glover, J., Conoley, J. and Witt, J. (eds.) *The Influence of Cognitive Psychology on Testing*. LEA, Hillsdale, New Jersey.

Short, R., Horn, J. and McArdle, J. 1984. Mathematical statistical model building in analysis of developmental data. In: Emde, R. and Harmon, R. (eds.) *Continuities and Discontinuities in Development*. Ablex, Norwood, New Jersey.

Siegler, R. 1981. Developmental sequences within and between concepts. *Monogr. Soc. Res. Child. Dev.* **46**(2), 84.

Spada, H. 1977. Logistic models of learning and thought. In: Spada, H. and Kempf, W. (eds.) *Structural Models of Thinking and Learning*. Huber, Bern.

Spada, H. and Kluwe, R. 1980. Two models of intellectual development and their reference to the theory of Piaget. In: Kluwe, R. and Spada, H. (eds.) *Developmental Models of Thinking*. Academic Press, New York.

Spada, H. and McGaw, B. 1985. The assessment of learning effects with linear logistic test models. In: Embretson, S. (ed.) *Test Design: Developments in Psychology and Education*. Academic Press, Orlando, Florida.

Tisak, J. 1984. *Exploratory longitudinal factor analysis in multiple populations with applications to growth in intelligence*. Unpublished doctoral dissertation, Department of Psychology, University of California, Berkeley, California.

Usiskin, Z. 1982. *Van Hiele levels and achievement in secondary school geometry.* CDASSG Project Report, University of Chicago, Chicago, Illinois.

van Hiele, P. M. 1986. *Structure and Insight: A Theory of Mathematics Education.* Academic Press, Orlando, Florida.

Wang, W., Wilson, M and Adams, R. 1996. Rasch Models for Multidimensionality Between and Within Items. In: Wilkson, M., Endelhard, G. and Draney, K. (eds.) *Objective Measurement IV: Theory into practice.* Ablex, Norwood, New Jersey.

White, R. T. 1988. *Learning Science.* Blackwell, Oxford.

Wilson, M. 1988. Detecting and interpreting local item dependence using a family of Rasch models. *Appl. Psychol. Meas.* **12**(4), 53–64.

Wilson, M. 1989a. Saltus: A psychometric model of discontinuity in development. *Psych. Bull.* **105**(2), 276–89.

Wilson, M. 1989b. Empirical examination of a learning hierarchy using an Item Response Theory model. *J. Exp. Educ.* **57**(4), 357–71.

Wilson, M. and Adams, R. 1995. Rasch models of item bundles. *Psychometri.* **60**(2), 181–98.

Wilson, M. and Draney, K. 1996. Partial credit in a developmental context: The case for adopting a mixture model approach. In: Wilson, M., Engelhard, G. and Dranct, K. (eds.) *Objective Measurement IV: Theory into practice.* Ablex, Norwood, New Jersey.

Wright, B. and Stone, M. 1979. *Best Test Design.* MESA, Chicago, Illinois.

Yamamoto, K. 1987. *A model that combines IRT and latent class models.* Unpublished doctoral dissertation, Department of Psychology, University of Illinois, Champaign-Urbana, Illinois.

Further reading

Adams, R. and Wilson, M. 1996. Formulating the Rasch model as a mixed coefficients multinomial logit. In: Engelhard, G. and Wilson, M. (eds.) 1996. *Objective Measurement: Theory into Practice III.* Ablex, Norwood, New Jersey.

13 Equating of Tests

M. J. Kolen

Multiple forms of educational tests contain different items, but are built to the same specifications. The forms typically differ somewhat in difficulty. Test form equating often adjusts scores on the multiple forms to account for these difficulty differences. This entry considers the procedures employed for the equating of tests.

1 Purpose and context

Multiple forms of a test built to the same content and statistical specifications, but containing different items, are often needed for test security purposes and for comparing examinee performance over time. Test forms typically differ somewhat in difficulty, even though test developers attempt to construct test forms to be as similar as possible to one another. Equating is often used when multiple forms of a test exist and examinees taking different forms are compared with one another. Equating makes adjustments for difficulty differences, thus allowing forms to be used interchangeably. After successful equating, examinees are expected to earn the same score regardless of the test form administered.

Equating has become a prominent topic in the United States for at least two reasons. First, there has been a growth of the type of testing that requires multiple forms. Second, test developers have had to make reference to equating in order to address issues raised by testing critics and test legislation. This increased visibility has led to various publications on equating. Before 1980, Angoff (1971) was the primary comprehensive source on equating. Since 1980, the Holland and Rubin (1982) book, the chapter by Petersen *et al.* (1989), and the Kolen and Brennan (1995) book have been published. In addition, introductory treatments of traditional equating methods by Kolen (1988) and item response theory methods by Cook and Eignor (1991) have been published. Also, there have been many journal articles, presentations of papers at meetings, and research reports. Papers by Cook and Petersen (1987), Skaggs and Lissitz (1986), a special issue of *Applied Psychological Measurement* (Brennan, 1987), and a special issue of *Applied Measurement in Education* (Dorans, 1990) provide an introduction to this literature. Because equating has been used predominantly with multiple-choice tests, the focus of this entry is on equating multiple-choice tests. The equating of other types of tests is discussed briefly.

Processes similar to equating exist that are better referred to as scaling to achieve comparability (Petersen *et al.*, 1989). One of these processes is vertical scaling (frequently referred to as vertical "equating"), which is used with elementary achievement test batteries. In these batteries, students are typically administered test questions matched to their current educational level (e.g., grade level). Vertical scaling allows the scores of examinees at different levels to be compared and allows for the assessment of an individual's growth over time.

Another example of scaling to achieve comparability is converting scores on one test to the score scale of another test, where both tests are measuring similar constructs, such as reading tests from two different publishers. As with vertical scaling, solutions to these problems do not allow the tests to be used interchangeably because the content of the tests is different. In addition, the development of these tests typically is not coordinated, so there are often differences in what the tests are measuring.

Although similar statistical procedures are often used in scaling and equating, their purposes are distinct. Whereas scaling is used to make scores comparable on tests that are purposefully built to be different, equating is used to make scores interchangeable on multiple forms that are built to be as similar as possible.

Scale scores are typically used in testing programs that use multiple forms. In these programs, a score scale is established based on the initial form of the test. The score scale is chosen to enhance the interpretability of scores by incorporating useful information into the score scale. The scale is maintained through an equating process that places scores from subsequent forms on the score scale that was established initially.

2   Properties of equating relationships

Define Form X as a new form and Form Y as an old form. Assume that a conversion of Form Y raw scores to scale scores was previously constructed, and that the purpose of equating is to be able to convert Form X raw scores first to Form Y raw scores and then to scale scores.

There are various desirable properties for equating relationships. For the "symmetry property" to hold, the function used to transform a score on Form X to the Form Y scale must be the inverse of the function used to transform a score on Form Y to the Form X scale. This symmetry property rules out regression as an equating method, because the regression of y on x is, in general, different from the regression of x on y. The "same-specifications property" requires that forms be built to the same content and statistical specifications. Symmetry and same-specifications are two properties that need to be achieved in order that scores across multiple test forms can be used interchangeably.

There are other properties of equating that are desirable, but that might not always be achievable such as "group invariance". Ideally, the equating relationship should be the same regardless of the group of examinees used to conduct the equating. Because this property cannot be expected to hold precisely for all populations of examinees, the population of examinees used to develop an equating relationship needs to be clearly stated.

The property of "equal distributions" states that score distributions should be the same across equated multiple forms for a given population. Define x as score on Form X and y as score on Form Y. Also, define F as the distribution function of equated x and G as the distribution function of y. If e is the function used to equate scores on Form X to the Form Y scale, then the equal distribution property is achieved if

$$F[e(x)] = G(y). \tag{1}$$

When the equal distribution property holds, the same proportion of examinees in the population would score below any particular cut score on any equated form.

The "equity" property, as described by Lord (1980), states that, after equating, examinees should be indifferent to which form they are administered. Precise specification of this definition of equity requires the consideration of a test theory model. Define $F[e(x)|\tau]$ as the cumulative distribution of converted scores for examinees with a particular true score, and $G(y|\tau)$ as the cumulative distribution on Form Y for examinees with a particular true score. Equity holds if

$$F[e(x)|\tau] = G(y|\tau) \text{ for all } \tau. \tag{2}$$

This definition implies that examinees with a given true score have identical observed score means, standard deviations, and distributional shapes of converted scores on Form X and scores on Form Y. Note that identical standard deviations implies that the standard error of measurement at any true score needs to be equal on the two forms. Lord (1980) showed that, under fairly general conditions, the equity property holds only if Form X and Form Y are essentially identical.

Morris (1982) suggested a less restrictive version of equity, referred to as "first-order equity". If first-order equity holds, examinees with a given true score have the same mean converted score on Form X as they have on Form Y. Defining E as the expectation operator, an equating achieves first-order equity if

$$E[e(x)|\tau] = E(y|\tau) \text{ for all } \tau. \tag{3}$$

The achievement of all the desirable properties is most likely impossible, although they might be closely approximated in actual equating situations. Typically, one or more of these properties are made the focus of a particular equating.

3   Equating designs

A variety of designs (Angoff, 1971) is used for collecting data for equating. The group of examinees included in the equating study should be reasonably representative of the group of examinees that will be given the test under typical test administration conditions. The choice of a design involves considering both practical and statistical issues. Three of the most commonly used designs that involve equating one intact test form with another such form are considered here. Pre-equating designs (Petersen *et al.*, 1989), in which the new form is constructed from items or parts of previously administered tests, are not considered here.

3.1   Random groups design

A spiraling process is typically used to implement the random groups design, where alternate examinees in a test center are administered the forms to be equated. In one method for spiraling, Form X and Form Y are alternated when the test booklets are packaged. When the booklets are handed out, the first examinee receives Form X, the second examinee Form Y, the third examinee Form X, and so on. If large groups of examinees are used that are representative of the group of examinees that will be given the test under typical test administration conditions, then the difference between group-level performance on the two forms is a direct indication of the difference in difficulty between the forms.

In the random groups design, each examinee takes only one form of the test, thus minimizing testing time. However, the random groups design requires the forms to be available and administered at the same time, which might be difficult in some situations. Also, because different examinees take the forms to be equated, larger sample sizes are needed than in some other designs.

3.2   Single group with counterbalance design

In the single group with counterbalancing design the same examinees are administered both Form Y and Form X. To deal with order effects, the order of administration of the two forms is counterbalanced. In one method for counterbalancing, one half of the test booklets are printed with Form X following Form Y, and the other half are printed with Form Y following Form X. In packaging, booklets having Form X first are alternated with booklets having Form Y first. When the booklets are handed out, the first examinee gets Form X first, the second examinee Form Y first, the third examinee Form X first, and so on. When the tests are administered, each test form is separately timed.

In the single group equating design with counterbalancing, the data from the form administered second are useful only if the order effects are constant across forms. For this reason, order effects need to be studied empirically. If order effects are not constant across forms, then the data from the form administered second might need to be disregarded.

Also, because two forms must be administered to the same students, testing time needs to be doubled, which often is not practically feasible. If fatigue and practice are effectively controlled by counterbalancing, then the single group design with counterbalancing has relatively small sample size requirements because, by taking both of the forms, each examinee serves as his or her own control.

3.3   Common-item nonequivalent groups design

The common-item nonequivalent groups design is used when more than one form per test date cannot be administered because of test security or other practical concerns. In this design, Form X and Form Y have a set of items in common, and different groups of examinees are administered the two forms. There are two variations of this design. When the score on the set of common items contributes to the examinee's score on the test, the set of common items is referred to as internal. Typically, internal common items are interspersed among the other items in the test. When the score on the set of common items does not contribute to the examinee's score on the test, the set of common items is referred to as external. Typically, external common items are administered as a separately timed section.

In this design, test score summary statistics are influenced by a combination of examinee group differences and test form differences. A major task in equating using this design is to disentangle these group and form differences.

The common-item nonequivalent groups design is widely used in practice. A major reason for its popularity is that this design requires that only one test form be administered per test date. The administrative flexibility offered by nonequivalent groups is gained at some cost. Strong statistical assumptions are required to disconfound group and form differences.

A variety of practical approaches were described by Brennan and Kolen (1987) to deal with the problems associated with this design. One important consideration is that the set of common items should be proportionally representative of the total test forms in content and statistical characteristics, as illustrated by Klein and Jarjoura (1985). That is, the common-item set should be a miniature version of the total test form, and needs to be a large proportion (say, at least 20 percent) of the total test. The common items also need to behave similarly on the old and new forms, as illustrated by Zwick (1991). To help ensure similar behavior, each common item should occupy a similar location on the two forms. Also, the common items should be exactly the same (e.g., no wording changes or rearranging of alternatives) in the old and new forms. A large number of common items should be used, as this usually allows for better content representativeness and greater stability. In addition, the groups administered the two forms should be reasonably similar, because extreme dissimilarities in groups have been found to cause problems in equating, as illustrated by Cook *et al*. (1988).

4   Equating relationships and estimation

To conduct equating, the relationship between alternate forms needs to be specified and estimated. The focus in traditional observed score equating methods is on the scores that are observed, although sometimes test theory models are used to estimate observed score equating relationships. Item response theory (IRT) and true score equating methods exist that make heavy use of test theory models.

4.1 Traditional equating methods

In traditional observed score equating methods, score correspondence is found by setting certain characteristics of the score distributions equal for a specified population of examinees. For example, in equipercentile equating, a score on Form X is considered to be equal to a score on Form Y if the two scores have identical percentile ranks in a population. Thus, x is equivalent to y if in the population of examinees.

$$F[e(x)] = G(y) \qquad (4)$$

The goal of equipercentile equating is for the distribution of converted scores on Form X to be the same as the distribution of scores on Form Y. This goal is the equal distribution property stated in Equation (1).

Linear equating is another observed score method of equating. The goal of linear equating is for the mean and standard deviation of converted scores to be equal on the two forms. The linear equating function in the population is

$$l(x) = \sigma(y) \left[\frac{x - \mu(x)}{\sigma(x)} \right] + \mu(y) \qquad (5)$$

where μ is a population mean and σ is a population standard deviation.

The estimation of equating relationships in the random groups design involves substituting statistics for the parameters in Equations 4 and 5. In equipercentile equating, the resulting equating relationship is typically irregular and subject to a considerable sampling error at those points where few examinees score. For this reason, smoothing methods are used. Methods for smoothing the score distributions are referred to as presmoothing methods (see Kolen, 1991, for a review) and methods for smoothing the equipercentile relationship as postsmoothing methods (e.g., Kolen, 1984). Either type of method, used analytically, can improve equating precision.

If the shapes of the score distributions for Form X and Form Y differ, which often occurs when the forms differ in difficulty, then equipercentile and linear relationships differ. However, even in these situations, the linear and equipercentile relationships are similar near the mean score. When interest is primarily in scores near the mean, such as when a cutting score is used that is near the mean score, then linear methods are sufficient. When interest is in scores all along the score scale and sample size is large, then equipercentile equating is often preferable to linear equating. A common rule of thumb is that a minimum of 1,000–2,000 examinees per form are needed for equipercentile equating, whereas fewer examinees are needed for linear equating.

The estimation of observed score equivalents in the common-item nonequivalent groups design requires that strong statistical assumptions be made. Define score on the common items as v. In the Tucker linear method, the linear regression of x on v is assumed to be equal for the examinees taking Form X and the examinees taking Form Y. A similar assumption is made about the linear regression of y on v. In the Levine linear observed score method, similar assumptions are made about true scores rather than observed scores. No method exists to directly test these assumptions using the data that are collected for equating, because examinees take only one form. An equipercentile counterpart of the Tucker method exists that is referred to as frequency estimation (Angoff, 1971).

Linear methods have been developed for equating true scores on test forms. The Levine true score method is based on assumptions that are similar to the Levine observed score method, but equates true scores. Hanson (1991) showed that applying the Levine true score equating relationship to observed scores results in the property of first-order equity when the test forms meet congeneric test theory model assumptions. Linear true score methods are likely to be most useful when tests differ in length, and therefore in reliability. For this reason, linear true score methods are often referred to as methods for unequally reliable tests. Curvilinear methods for equating true scores exist that make use of strong true score models (see Lord, 1980, Chapter 17).

4.2 IRT methods

Item response theory (IRT) models assume that examinee ability (θ) can be described by a single latent variable and that items can be described by a set of item parameters (Lord, 1980). For multiple-choice tests, the probability that examinees of ability θ correctly answer item g is symbolized $P_g(\theta)$. Item response theory models are based on strong statistical assumptions that need to be checked before the methods are applied. The θ-scale has an indeterminate location and spread. For this reason, one θ-scale sometimes needs to be converted to another linearly related θ-scale. If scores are to be reported on the number-correct scale, then there are two steps in IRT equating. First, the θ-scales for the two forms are considered to be equal or set equal. Then number-correct score equivalents on the two forms are found.

In many situations, the parameter estimates for the two forms are on the same θ-scale without further transformation. In general, no transformation of the θ-scales is needed in the following situations: (a) in the random groups design or the single group design when the parameters for Form X and Form Y are estimated separately using ability scales with the same mean and standard deviation; (b) in the single group design when parameters for the two forms are estimated together; and (c) in the common-item nonequivalent groups design when Form X and Form Y parameters are estimated simultaneously. The typical situation in which a transformation of the θ-scale is required is in the common-item nonequivalent groups design when the Form X and Form Y parameters are estimated separately. In this situation, the parameters for one form are linearly transformed to the θ-scale of the other form. In one procedure for equating, the mean and standard deviation of the estimated θ-values are set equal. Additional methods of equating the θ-scales also exist (see Haebara, 1980; Stocking and Lord, 1983).

After the parameter estimates are on the same scale, IRT true score equating can be used to relate number-correct scores on Form X and Form Y for score reporting purposes. In this procedure, the true score on one form associated with a given θ is considered to be equivalent to the true score on another form associated with that θ. In IRT the true score on Form X for an examinee of ability θ is defined as

$$\tau_X(\theta) = \sum_{g:X} P_g(\theta) \tag{6}$$

where the summation $g:X$ is over items on Form X. True score on Form Y for an examinee of ability θ is defined as

$$\tau_Y(\theta) = \sum_{g:Y} P_g(\theta) \tag{7}$$

where the summation $g:Y$ is over items on Form Y. Given a θ, the score on Form X, $\tau_X(\theta)$, is considered to be equivalent to the score on Form Y, $\tau_Y(\theta)$. In practice, estimates of the parameters are used in Equations (6) and (7).

As a practical procedure the estimated true score relationship is applied to observed scores. However, no theoretical reason exists for treating scores in this way. Rather, doing so has been justified in item response theory by showing that the resulting true score conversions are similar to observed score conversions (Lord and Wingersky, 1984). Procedures exist for using IRT methods to conduct equipercentile equating (Lord, 1980), although these methods have not been used much in practice.

Any application of unidimensional IRT methods requires that the test forms be unidimensional and that the relationship between ability and probability of correct response follow a specified model. These requirements are difficult to justify for many educational achievement tests, although the methodology might be robust to violations in some circumstances. Item response theory methods also have substantial sample size requirements, although the sample sizes for the Rasch model (Wright and Stone, 1979) are considerably less than those for the three parameter logistic model (Lord, 1980). In any application of IRT to equating, the fit of the models needs to be carefully analyzed.

5   Equating issues

Minimizing equating error is a major goal when deciding whether or not to equate forms, when designing equating studies, and when conducting equating. Random equating error is present whenever samples from populations of examinees are employed to estimate parameters (e.g., means and standard deviations) that are used to estimate an equating relationship. Random error diminishes as sample size increases. Random error is typically indexed by the standard error of equating.

Systematic equating error results from violations of the assumptions of the particular equating methodology used. For example, in the common-item nonequivalent groups design, systematic error results if the assumptions of statistical methods used to disconfound form and group differences are not met. Systematic error might also be present if the group of examinees used to conduct the equating differs substantially from the group of examinees who are administered the test operationally. Systematic equating error typically cannot be quantified in actual equating situations.

Equating error of both types needs to be controlled because it can propagate over-equatings and result in scores on later forms not being comparable to scores on earlier forms. In testing programs that use many test forms, the selection of the previously equated form or forms influences equating error. For example, in the common-item nonequivalent groups design a new test form is often equated to two previously equated forms by including two sets of common items. The results from each of the two equatings could be compared for consistency, and this provides a check on the adequacy of the equating. The results could be averaged, which might provide more stable equating.

Many testing programs, because of practical constraints, cannot use equating for comparing examinee scores and assessing score trends over time. For example, sometimes security issues prevent test items from being administered more than once, such as when all items administered are released after the test is given. All equating designs require more than one administration of at least some items. In other practical situations the amount of equating error might be so large that equating would add more error into the scores than if no equating had been done. For example, suppose that the common items behave differently from one testing to the next because of changes in item position. In such a case, no equating might be preferable to equating, using the common items. Equating with small sample sizes might also introduce more error than it would remove. In these situations, no equating might be preferable to equating. Even without equating, scores of examinees who are given the same form can be compared with one another. If most decisions are made among examinees tested with the same form, then not conducting equating might have minimal consequences. However, not equating would make it difficult to chart trends over time and to compare examinees who were given different forms.

In some circumstances, equating can be used with nonmultiple-choice tests. To equate such tests with observed-score methods, equating proceeds in the same way as it does with multiple-choice tests. With IRT, if the test items are not dichotomously scored, then generalizations of the standard IRT models must be used, such as the model developed by Masters (1982). One practical problem often occurs when equating tests that are not multiple-choice with the common-item nonequivalent groups design. In this situation, it can be very difficult to develop a set of common items that is representative of the entire test, because there are so few items on the test. This problem is especially severe with essay tests, where a test form might contain only one or two essay items (see *Essays: Equating of Marks; Partial Credit Models; Rating Scale Analysis*). Recently there has also been considerable interest in developing comparable scores between computerized tests and paper-and-pencil tests, and between alternate forms of computerized tests. Kolen and Brennan (1995) review many of the related issues.

6 Equating and related procedures in testing programs

To provide an indication of the variety of equating procedures that are in use, some examples of equating methods are presented. In the United States, the ACT (American College Testing, 1989) and the SAT (Scholastic Aptitude Test) (Donlon, 1984) are the two major tests used for college admissions purposes. Different forms of the tests are given multiple times during any year. Examinees who test on different dates are compared to one another, so there is a need to conduct equating.

The ACT tests are equated using a random groups design. Forms are spiraled in test centers in a special operational equating administration. The forms are equated using equipercentile methods with postsmoothing. The equated forms are used in subsequent administrations.

The SAT is equated on each test date using a common-item nonequivalent groups design with external common items. Typically, a single form is administered along with an equating section that is not part of the score of the examinee, but the examinees do not known which section is not scored. The use of the external common-item section allows the scored items of the examination to be released to examinees while still allowing for equating in the future. The forms are equated to two previously administered forms using two common-item sections. When the SAT is equated, IRT, linear, and equipercentile methods are all examined.

The Iowa Tests of Basic Skills (Hieronymus and Hoover, 1986) are vertically scaled using a scaling test method, in which a scaling test is constructed to span various levels of educational development. The scaling test is spiraled with the tests that are typically used at each educational level. Equipercentile procedures are used to scale each test to the scaling test. The California Achievement Tests are vertically scaled by using IRT methods. Yen (1983) provided a detailed discussion of these methods. Petersen et al. (1989) discussed issues in vertical scaling.

The National Assessment of Educational Progress (NAEP) in the United States uses common-item methods with item response theory methods to conduct equating (Zwick, 1991). This program also uses item sampling to estimate group-level achievement rather than achievement at the individual level. Intriguing scaling issues in NAEP arise from changes in specifications over time.

In a study to examine the feasibility of using the Rasch latent trait measurement model in the equating of the Australian Scholastic Aptitude Test, Morgan (1982) found that all items used in the test needed to fit a unidimensional latent trait model at the test development stage. On two occasions, in 1970–71 and 1983–84, Keeves and Schleicher (1992) also used the Rasch model to equate and scale science achievement tests employed at the 10-year-old, 14-year-old, and secondary school levels in 10 different countries through the use of different sets of bridging items.

7 Conclusion

The equating methods outlined in this entry have been shown to produce adequate equating results for the random groups design when the following conditions are met:

(a) Test forms are built to the same carefully defined content and statistical specifications.
(b) The examinee groups are reasonably representative of the group of examinees who are going to be administered the test operationally.

In addition to (a) and (b), the common-item nonequivalent groups design requires additional conditions:

(c) The groups taking the two forms are similar to one another.
(d) The common items are representative of the full test in content and statistical characteristics, and are at least 20 percent of the full test.
(e) The common items behave similarly in the forms that are equated.

Item response theory methods also require that the test forms are at least close enough to being unidimensional for practical purposes. Linear methods also require either that the score distributions be similar in shape or that interest is only in scores near the mean score. Equipercentile and IRT methods require large sample sizes.

The adequacy of scaling to achieve comparability and vertical scaling depends on the purpose of the scaling and how the results are to be used. Scaling to achieve comparability is probably much more dependent on the groups sampled and on the design used to collect the data, than is equating (see also: *Item Response Theory; Latent Trait Measurement Models*).

References

American College Testing. 1989. *Preliminary Technical Manual for the Enhanced ACT Assessment.* American College Testing, Iowa City, Iowa.
Angoff, W. A. 1971. Scales, norms, and equivalent scores. In: Thorndike, R. L. (ed.) 1971. *Educational Measurement*, 2nd edn. American Council on Education, Washington, DC.
Brennan, R. L. (ed.) 1987. Problem, perspectives and practical issues in equating. *Appl. Psychol. Meas.* **11**(3), 221–306.
Brennan, R. L. and Kolen, M. J. 1987. Some practical issues in equating. *Appl. Psychol. Meas.* **11**(3), 279–90.
Cook, L. L. and Eignor, D. R. 1991. NCME instructional module: IRT equating methods. *Educational Measurement: Issues and Practice* **10**(3), 37–45.
Cook, L. L., Eignor, D. R. and Taft, H. L. 1988. A comparative study of the effects of recency of instruction on the stability of IRT and conventional item parameter estimates. *J. Educ. Meas.* **25**(1), 31–45.
Cook, L. L. and Petersen, N. S. 1987. Problems related to the use of conventional and item response theory equating methods in less than optimal circumstances. *Appl. Psychol. Meas.* **11**(3), 225–44.

Donlong, T. F. (ed.) 1984. *The College Board Technical Handbook for the Scholastic Aptitude Test and Achievement Tests*. College Entrance Examination Board, New York.

Dorans, N. J. 1990. Equating methods and sampling designs. *Applied Measurement in Education* **3**(1), 3–17.

Hanson, B. A. 1991. A note on Levine's formula for equating unequally reliable tests using data from the common item nonequivalent groups design. *J. Ed. Stat.* **16**(2), 93–100.

Haebara, T. 1980. Equating logistic ability scales by a weighted least squares method. *Japanese Psychological Research* **22**(3), 144–9.

Holland, P. W. and Rubin, D. B. (eds.) 1982. *Test Equating*. Academic Press, New York.

Hieronymus, A. N. and Hoover, H. D. 1986. *Iowa Tests of Basic Skills Manual for School Administrators Levels 5-14 ITBS Forms G/H*. Riverside Publishing Co., Chicago, Illinois.

Keeves, J. P. and Schleicher, A. 1992. Changes in science achievement: 1970–1984. In: Keeves, J. P. (ed.) *The IEA study in Science III. Changes in Science Education and Achievement: 1970–1984*. Pergamon Press, Oxford. pp. 263–290.

Klein, L. W. Jarjoura, D. 1985. The importance of content representation for common-item equating with nonrandom groups. *J. Educ. Meas.* **22**(3), 197–206.

Kolen, M. J. 1984. Effectiveness of analytic smoothing in equipercentile equating. *J. Ed. Stat.* **9**(1), 25–44.

Kolen, M. J. 1988. An NCME instructional module on traditional equating methodology. *Educational Measurement: Issues and Practice* **7**(4), 29–36.

Kolen, M. J. 1991. Smoothing methods for estimating test score distributions. *J. Educ. Meas.* **28**(3), 257–82.

Kolen, M. J. and Brennan, R. L. 1995. *Test Equating. Methods and Practices*. Springer-Verlag, New York.

Lord, F. M. 1980. *Applications of Item Response Theory to Practical Testing Problems*. Erlbaum, Hillsdale, New Jersey.

Lord, F. M. and Wingersky, M. S. 1984. Comparison of IRT true-score and equipercentile observed-score "equatings". *Appl. Psychol. Meas.* **8**(4), 453–61.

Masters, G. N. 1982. A Rasch model for partial credit scoring. *Psycholmetri.* **47**(2), 149–74.

Morgan, G. 1982. The use of the Rasch latent trait measurement model in the equating of Scholastic Aptitude Tests. In: Spearritt, D. (ed.) 1982. *The Improvement of Measurement in Education and Psychology*. ACER, Hawthorn. pp. 189–212.

Morris, C. N. 1982. On the foundations of test equating. In: Holland, P. W. and Rubin, D. B. (eds.) 1982. *Test Equating*. Academic Press, New York.

Petersen, N. S., Kolen, M. J. and Hoover, H. D. 1989. Scaling, norming, and equating. In: Linn, R. L. (ed.) 1989. *Educational Measurement*, 3rd edn. American Council on Education and Macmillan, New York.

Skaggs, G. and Lissitz, R. W. 1986. IRT test equating: Relevant issues and a review of recent research. *Rev. Educ. Res.* **56**(4), 495–529.

Stocking, M. L. and Lord, F. M. 1983. Developing a common metric in item response theory. *Appl. Psychol. Meas.* **7**(2), 201–10.

Wright, B. D. and Stone, M. H. 1979. *Best Test Design*. Mesa Press, Chicago, Illinois.

Yen, W. 1983. Use of the three-parameter model in the development of a standardized test. In: Hambleton, R. L. (ed.) 1983. *Applications of Item Response Theory*. Educational Research Institute of British Columbia, Vancouver.

Zwick, R. 1991. Effects of item order and context on estimation of NAEP reading proficiency. *Educational Measurement: Issues and Practice* **10**(3), 10–16.

14 Essays: Equating of Marks

D. Andrich

The science of educational measurement has developed in order to enhance the validity and reliability of tests. Unfortunately, reliability can be attained readily at the expense of validity, which is illustrated by the use of tests composed of multiple-choice items compared to those that require extended responses. The multiple-choice item was developed because of its reliability in scoring, and has been accompanied by a voluminous amount of methodological research according to classical test theory (CTT), general-izability test theory (GTT), and latent trait theory (LTT). One feature of LTT is that it uses a response function with the item as the unit of analysis, making it possible to use different items for different persons, yet still estimate the location of all persons and all items on the same continuum (Weiss, 1983; Julian and Wright, 1988; Kubinger, 1988).

Despite this theoretical and practical work on multiple-choice items, schools, colleges, and universities continue to require students to provide extended responses, often in the form of essays. Essays are considered a more valid means for assessing the higher levels of knowledge associated with analyzing, synthesizing and evaluating information (e.g., Bloom *et al.*, 1956).

Before proceeding, an important point of nomenclature must be clarified. It is common in LTT to refer to the location of a person or a trait as the person's *"ability"*. This term can sometimes be interpreted as some ability determined entirely genetically. However, that is not the intention in LTT: instead, it is intended that a latent or underlying ability to produce a set of related manifest performances is inferred from the manifest performances. This latent ability to perform arises from the sum of all the person's relevant learning experiences. To avoid misinterpretation, the term *"ability/performance"* will be used throughout for the location on the latent trait.

Surprisingly, relatively little research on LTT has been applied to essays where the scores may range beyond single digit integers even though many features are directly analogous. First, more than one question is usually to be answered, which has the same purpose as having many multiple-choice questions – to sample a greater variety of content and to have a more precise estimate of each person's ability/performance. Second, often there is a choice, which acknowledges that it is not necessary to answer only one set of questions to be able to provide evidence of appropriate knowledge in a certain field. Traditional text books (e.g., Chase, 1978) consider that choice should not be permitted because scores obtained on different combinations of questions cannot be compared properly. Nevertheless, choice is provided in essay-type examinations, in part because of time constraints, but in part because it is considered more valid to provide students with some choice. The task then is to account for the difficulties of the questions, which is again analogous to the multiple-choice context–items are invariably of different difficulty, and using LTT it is possible to locate persons on the same continuum even though they have answered different questions. This entry applies a latent trait model to equate questions of an examination which was composed of four questions, where the maximum mark for each question was 25, and where the students were to answer any two of the four questions.

1   The model

The model applied is an extension of Rasch's (1980) simple logistic model (SLM) for dichotomously scored items (Andrich, 1988; Wright and Stone, 1979). This model was derived theoretically by Rasch (1968), Andersen (1977) and Andrich (1978), and has been studied further (Wright and Masters, 1982; Andrich, 1985; Jansen and Roskam, 1986), usually where the number of graded responses has been limited to single digit values. However, this procedure has had very restricted application in practice.

The SLM specifies a location parameter (ability/performance β) of the person and a location parameter (difficulty δ) of the item and takes the form:

$$\Pr\{X_{ni}=x; \beta_n, \delta_i\} = [\exp\{x\beta_n - x\delta_i\}]/\gamma_{ni} \tag{1}$$

where $\gamma_{ni} = 1 + \exp\{\beta_n - \delta_i\}$ is the normalizing factor that ensures that $\Pr\{X_{ni}=0\} + \Pr\{X_{ni}=1\} = 1$. The extended logistic model (ELM) for ordered categories takes the form

$$\Pr\{x; \beta_n, \delta_i, \tau_i\} = [\exp\{x(\beta_n - \delta_i) - \tau_{1i} - \tau_{2i} \ldots - \tau_{xi}\}]/\gamma_{ni} \tag{2}$$

where (a) τ_{xi} is the threshold between the $(x-1)$th and the xth response categories, there being m thresholds with $\tau_0 \equiv 0$; (b) $x \in \{0, 1, 2, \ldots, x, \ldots, m\}$ corresponds to the ordered category of the response; and (c)

$$\gamma_{ni} = \sum_{k=1}^{m} \exp\left[k(\beta_n - \delta_i) - \left(\sum_{y=1}^{k} \tau_{yi}\right)\right]$$

again is the normalizing factor. With m thresholds, a constraint such as $\sum_{x=1}^{m} \hat{\tau}_{xi}=0$ is necessary within each item and $\sum_{i=1}^{I} \hat{\delta}_i=0$ across items. The curve of the rate of change of the expected value (EVC) with respect to β is given by

D. Andrich

$$E[X_i; \beta, \delta_i, \tau_i] = \sum_{x=1}^{m} x \Pr\{x; \beta, \delta_i, \tau_i\}. \tag{3}$$

These curves are shown in Figure 1 for each question of the example. It can be seen that they have different locations (difficulties) and slopes. Different slopes result from different distances between the thresholds among questions and they correspond to the traditional

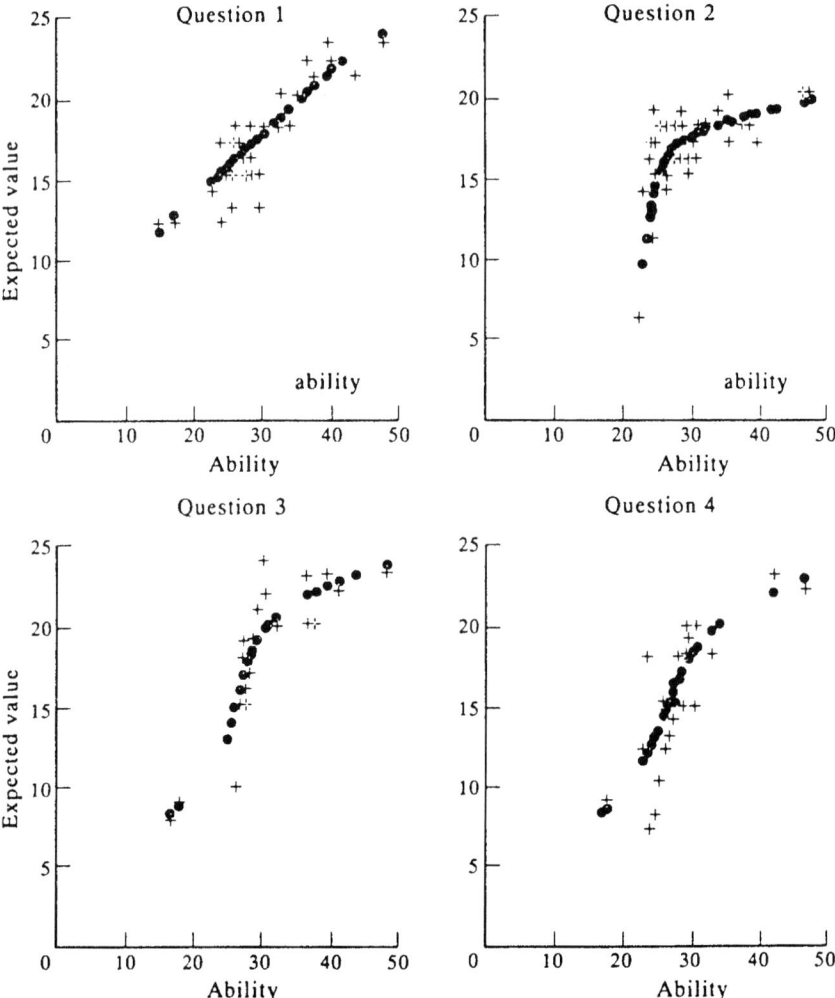

Figure 1   Points on the expected value curve (●) and observed scores (+) for each question.

notion of discrimination – the closer the thresholds, the greater the discrimination.

It is possible to estimate all thresholds for each question. However, where the maximum score is large, for example $m = 25$ as in the data set in this entry, it is extravagant to estimate all parameters. Here, the thresholds are reparameterized so that their spread, skew and kurtosis are summarized: specifically,

$$-\sum_{k=1}^{x} \tau_{ki} = x_i(m_i - x_i)\theta_i + x_i(m_i - x_i)(2x_i - m_i)\eta_i + x_i(m_i - x_i)(5x_i^2 - 5x_i m_i + m_i^2 + 1)\psi_i$$

where θ_i is the average equal half-distance between thresholds, and where η_i and ψ_i indicate the asymmetric deviation (skewness) and symmetric deviation (kurtosis) from equidistance respectively. Pedler (1987) has derived the general expression for generating the successive polynomials for the category coefficients.

The following definitions further simplify the expressions: let

$$g_{1i} = g_1(x_i) = -x_i/m_i, \qquad\qquad\qquad \varphi(m_i) = \varphi_{1i};$$

$$g_{2i} = g_2(x_i) = x_i(m_i - x_i)/m_i^2, \qquad\qquad \varphi(m_i^2\theta_i) = \varphi_{2i};$$

$$g_{3i} = g_3(x_i) = x_i(m_i - x_i)(2x_i - m_i)/m_i^3, \qquad \varphi(m_i^3\eta_i) = \varphi_{3i};$$

$$g_{4i} = g_4(x_i) = x_i(m_i - x_i)(5x_i^2 - 5x_i m_i + m_i^2 + 1)/m_i^4, \quad \varphi(m_i^4\psi_i) = \varphi_{4i}.$$

The division of successive coefficients by m_i, m_i^2, etc., is used to account for the large value of m_i, and the consequent extremely small values of the parameters. Accordingly, the parameters δ_i, θ_i and η_i are rescaled so that $\delta_i \to m_i\delta_i$, $\theta_i \to m_i^2$, $\eta \to m_i^3 \eta_i$, and $\psi \to m_i^4\psi$. In the subsequent presentation and in the results section these rescaled parameters are assumed. Using vector notation, the final probability is given by the simple equation:

$$\Pr\{x_{ni}; \boldsymbol{\varphi}, \boldsymbol{\beta}_n\} = [\exp\{\mathbf{g}_i'\boldsymbol{\varphi}_i + x_{ni}\boldsymbol{\beta}_n\}]/\gamma_{ni}. \qquad (4)$$

The basis for estimating the item parameter is the probability of the response pair (x_{ni}, x_{nj}) of each person n to a pair of questions i and j conditional on the total score $r_{nij} = x_{ni} + x_{nj}$. This is given by

$$\Pr\{(x_{ni}, x_{nj})\,|\,r_{nij}; \boldsymbol{\varphi}_i, \boldsymbol{\varphi}_j\} = \frac{\exp[\mathbf{g}_i'\boldsymbol{\varphi}_i + \mathbf{g}_j'\boldsymbol{\varphi}_j]}{\displaystyle\sum_{(x_{ni}, x_{nj})\,|\,r_{nij}} \exp[\mathbf{g}_i'\boldsymbol{\varphi}_i + \mathbf{g}_j'\boldsymbol{\varphi}_j]} \qquad (5)$$

in which the terms

$$\frac{1}{\gamma_{ni}\,\gamma_{nj}} \quad\text{and}\quad \exp[_{nij}\boldsymbol{\beta}_n]$$

are eliminated.

This equation may be generalized across all pairs of items for each person, and then across all persons in deriving estimation equations for the item parameter estimates. Three points need to be noted in Equation (5). First, the person parameter β_n does not appear in the equation. Thus the estimates of the item parameters are independent of the distribution of abilities/performances. Equally importantly, the equation implies that for a given total score of a person on two items which are intended to reflect the same latent trait, the differences in the responses to these two items is a function of the properties of the items, and not of the person. These properties include the relative difficulties of the items as in the dichotomous case, and in addition, the tendency for the items to spread, skew and flatten responses of each person, independently of the abilities/performances of the different persons involved.

Second, when a person has a minimum total score of 0 on both items or a maximum score of $m_i + m_j$, the pair of responses are not included in the solution equations. Likewise, when a person has no response to an item, that item is not paired with any other item. Thus, missing data are handled routinely. Third, the generalization of Equation (5) across persons and items in the form of a continued product provides maximum likelihood-type estimates. However, because the same responses are used across different pairs of items, dependencies are created, and therefore the tests of fit and standard errors cannot be derived directly. The parameter estimates themselves are not affected unduly because the dependencies simply provide redundant, not contradictory, information. Nevertheless, further research, both analytic and through simulation studies following on from the work of van der Linden and Eggen (1986) in the dichotomous case, is required, and this is currently being carried out. This pairwise algorithm was studied extensively by Choppin (1983).

Given the estimates of the item parameters, and treating them as fixed, the ability/performance of each person is estimated by a direct maximum likelihood equation which takes the simple form

$$r_n = \sum_i x_{ni} \Pr\{x_{ni}; \hat{\beta}_n, \hat{\phi}\},\tag{6}$$

where

$$r_n = \sum_i x_{ni},$$

with the summations carried over the items to which person n has responded.

Various tests of fit can be devised to help check the degree to which the data conform to the model. In this case one test of fit used is a graphical one in which the observed value of each person on each of the questions is compared with its expected value given the estimate of ability/performance of each person and the estimate of the item parameters. A second one considers the standardized residuals for each person. These are illustrated in the context of the example.

2   The example

The example (see Table 1) involved a unit of study in social research methods at Murdoch University (Australia), and part of the assessment involved an examination in which students had to answer two of the four questions set. The possible range of marks for each question was 0 to 25. The examiners were experienced lecturers in the course, and they were all involved in the construction of the four questions. In addition, they discussed their marking schedules and agreed on them before they began. From a statistical point of view, the example involves a very small data set – only 72 persons responding to only two of four questions. However, its advantages are that it is a real example (already mentioned), and in being small, it permits the complete data set to be studied in detail to illustrate the significant features of the model, the analysis, and the interpretation. It also shows that large samples are not required to carry out a specific analysis.

In the assessment of the responses of the students, a different examiner graded each question. Thus it is impossible from this design to distinguish between the effects of examiners and the properties of the questions. However, the same analysis, with corresponding changes to the interpretation, could be carried out if the data had been generated by one grader who graded all questions, or if the data came from one question having been graded by four different graders.

It is clear that calculating the mean and other statistics for each question makes it impossible to compare them – the students chose the questions, the questions were not assigned at random to the students, and so it cannot be assumed that random variation would ensure that the average ability/performance of students answering each question was the same. Table 2 shows the estimates of the parameters obtained for each question. It is evident from this Table that there are some differences in location (difficulty) among the questions, with question 2 (Q2) being the most difficult and question 3 (Q3) the easiest. However, in the presence of different discriminations of the questions it is not straightforward to interpret differences in difficulty.

Figure 1 shows the expected value curves (EVCs), defined in Equation (3) for each question. In order to make the abilities have more customary values, the estimated abilities/performances were transformed linearly by multiplying them by 0.5 and adding 20: they are notated by $\hat{\beta}^* = 0.5\hat{\beta} + 20$. These figures show that the observed values tend to congregate around the EVCs. In addition, the curves for Questions 2 and 3 are somewhat similar: the other two are different from these and from each other. In traditional terms, Question 2 has not operated as effectively in discriminating among the students as the other questions. However, it is important to appreciate that unlike the typical situation with many multiple-choice items, this question cannot be eliminated. This point is elaborated in the discussion.

The six test characteristic curves (TCC) for each combination of two questions, are of major interest. Figure 2 shows these curves. It will be recalled that for each combination of questions, the total score is sufficient for estimating the ability/performance; therefore,

Table 1 Data from four exam questions, each with a maximum score of 25, in which only two questions needed to be answered: persons ordered by total score

Item			Q1 25[a]	Q2 25	Q3 25	Q4 25
Person	Total score	Ability/Performance	Q1	Q2	Q3	Q4
1	17	16.77			9	8
2	20	15.12	12			8
3	21	17.48	12			9
4	22	23.49		15		7
5	24	24.01		6	18	
6	26	23.07	14			12
7	26	24.55		18		8
8	27	24.84		14		13
9	28	23.99	15		13	
10	28	24.81		15	13	
11	28	25.15		18		10
12	28	25.15		15		13
13	29	24.66	17	12		
14	29	24.66	12	17		
15	30	25.39		20	10	
16	30	26.11	15			15
17	31	26.35		19		12
18	31	25.48	15	16		
19	31	26.35		19		12
20	32	26.04	13	19		
21	32	26.40	17		15	
22	32	26.88		19		13
23	32	27.56	17			15
24	32	27.56	17			15
25	33	26.77	17	16		
26	33	27.01	18		15	
27	33	27.01	17		16	
28	33	27.49		19		14
29	33	28.33	15			18
30	34	27.03		15	19	
31	34	27.03		16	18	
32	34	27.67	17		17	
33	34	28.21		19		15
34	35	28.40	16		19	
35	35	29.06		17		18
36	35	29.22	18	17		
37	35	29.22	15	20		
38	35	29.22	16	19		
39	35	29.22	15	20		
40	35	30.06	17			18
41	36	28.29		18	18	
42	36	28.29		17	19	
43	36	29.23	15		21	
44	36	30.08		17		19
45	36	30.08		16		20
46	36	30.90	18	18		
47	37	30.20	13		24	
48	37	30.20	15		22	
49	37	31.32		17		20
50	37	32.71	18	19		
51	38	30.61			20	18
52	38	33.42	20			18
53	38	34.57	18	20		
54	38	34.57	19	19		
55	39	31.71		19	20	
56	39	34.82		19		20
57	39	36.47	20	19		
58	40	38.42	21	19		
59	40	38.42	21	19		
60	41	36.22		18	23	
61	41	36.22		21	20	
62	41	40.46	23	18		
63	42	37.22	22		20	
64	42	37.22	20		22	
65	42	39.41		19	23	
66	42	42.67	22	20		
67	43	43.30		20	23	
68	43	47.66		21		22
69	44	40.83	22		22	
70	44	43.09	21			23
71	44	48.02		21	23	
72	44	48.44	23	21		

[a] Maximum score for each question is 25.0.

Table 2 Parameter estimates for each question: constraint on the location parameter only

	Location		Scale		Skew		Kurtosis	
	δ_i	(SE)	θ_i	(SE)	η_i	(SE)	ψ_i	(SE)
Q1	0.61	(1.82)	49.84	(4.05)	13.35	(5.61)	− 10.56	(5.90)
Q2	25.75	(1.15)	69.94	(4.74)	50.24	(3.59)	75.98	(5.74)
Q3	− 20.50	(1.64)	102.82	(4.19)	− 66.96	(5.09)	64.85	(6.37)
Q4	− 5.86	(1.54)	86.59	(5.41)	− 35.50	(4.22)	53.13	(6.33)
	0.00							

each total score for each combination transforms to a single ability/performance. As expected from the EVCs, the TCCs are also different from each other, so that the same total score on different combinations leads to different abilities. Interestingly, also, is that only two persons answered the combination of Questions 3 and 4, yet these total raw scores are mapped onto the same scale as are those of the other persons.

3 Further issues, interpretation and generalizations

So far, analogous features between tests composed of multiple-choice items and essay questions have been emphasized. There are, however, some important differences between the two, which rest primarily on the likely substantive applications of the procedure.

First, in university or college examinations it is unlikely that an examiner would wish to identify parameters of a question and use them again. Clearly, the questions may be

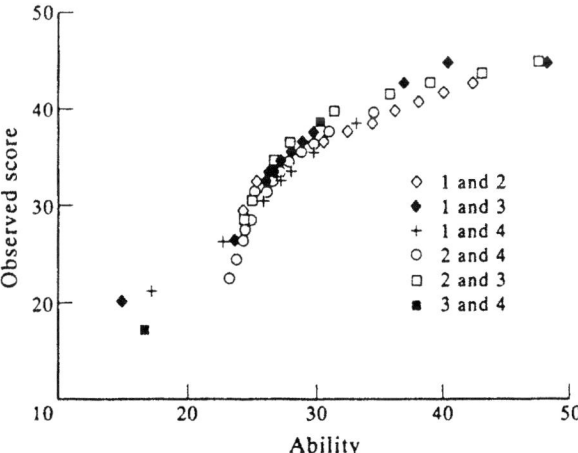

Figure 2 Points on the test characteristic for each of the combinations of questions mapped on to the same ability continuum.

retained or improved, but with the effect of the grader, the teaching, and so on it is more likely that the analysis would be repeated. Second, the transformed scores give a different ranking of students from the raw scores, even though the correlation in this case is high (0.92). This can have implications for who receives prizes, grades, or places in highly selective programs of study. Third, many of the usual aspects of programs of the test of fit do not apply. For example, even though Question 2 did not operate as well as it might have and as well as the others, there was no opportunity to discard the question – given time, a remarking of all answers to that question might have been possible, but, as usual, there was no time in this case, and the original data had to be used. However, if for some reason the usual formal test of fit between the data and the model, which focuses on the questions, was required, then the usual chi-square and likelihood ratio tests, which check specific violations of the model, can be used (Wright and Masters, 1982).

Fourth, therefore, the important test of fit is at the level of the individual. Recall that because the total score on the questions answered is sufficient for the ability/performance estimate, all persons with the *same total score on the same questions* will have the same ability/performance estimate. This implies that the pattern or profile of scores is immaterial, and that there is no further information in the profile. This indeed is the case, provided that the data of the profile accord with the model. However, if the profile does not accord with the model, then the total score cannot be interpreted as sufficient, and therefore there is information in the profile. The key index for further information compares the observed and expected values for each question given the parameter estimates. If these are similar, then the profile has been recovered from the total score. Of course, the recovery will not be perfect, and the decision as to the adequacy of the recovery will be the usual combination of statistical inference and an understanding of what might be an important substantive difference.

For example, consider the information on the two profiles in Table 3. It is clear from the size of the standardized residuals that the profile of person 17 is recovered better than that of person 5. Therefore, if there were the resources to remark some scripts, and if the choice were between reassessing the scripts of person 17 and person 5, the script of person 5 would be the one reassessed.

Table 3 Analysis of profiles of two persons who have answered the same two questions

Person	Questions	Ability/Performance estimate	Observed score	Expected score	Standardized redidual z	Σz^2
	2		19	16.563	0.683	
17		26.350				0.923
	4		12	14.435	− 0.676	
	2		6	11.901	− 0.974	
5		24.010				4.026
	4		18	12.099	1.754	

A single index for comparing consistency of profiles is given by the sum of the squared residuals, $\Sigma_i z_{ni}^2$, also shown in Table 3. Further research on the statistical properties of such an index is required, and while that research is important, the index can be used as it is in situations such as the one described. Already, many examining panels look for inconsistencies in marks awarded, and the procedure outlined systematizes this procedure by ordering the magnitude of the discrepancies. Then the number of profiles re-examined is as much a function of the expectations of examiners who have a substantive knowledge of the field, and the availability of resources, as it is a function of any statistical distribution (see also: *Rasch Measurement Theory; Rating Scale Analysis; Rasch Measurement Models*).

References

Andersen, E. B. 1977. Sufficient statistics and latent trait models. *Psychometri.* **42**, 69–81.

Andrich, D. 1978. A rating formulation for ordered response categories. *Psychometri.* **43**, 561–73.

Andrich, D. 1985. An elaboration of Guttman scaling with Rasch models for measurement. In: Brandon-Tuma, N. (ed.) 1985. *Sociological Methodology.* Jossey-Bass, San Francisco, California.

Andrich, D. 1988. *Rasch Models for Measurement.* Sage, Beverly Hills, California.

Bloom, B. S., Krathwohl, D. and Madaus, G. 1956. *Taxonomy of Educational Objectives.* McKay, New York.

Chase, C. L. 1978. *Measurement for Educations Evaluation*, 2nd edn. Addison-Wesley, Reading, Massachusetts.

Choppin, B., 1983. *A Fully Conditional Estimation Procedure for Rasch Model Parameters.* Center for the Study of Evaluation, Graduate School of Education, University of California, Los Angeles, California.

Jansen, P. G. W. and Roskam, E. E. 1986. Latent trait models and dichotomization of graded responses. *Psychometri.* **51**, 149–74.

Julian, E. R. and Wright, B. D. Using computerized patient simulations to measure the clinical competence of physicians. *Applied Measurement in Education* **1**, 299–318.

Kubinger, K. D. 1988. On a Rasch-model based test for noncomputerised adaptive testing. In: Langeheine, R. and Rost, J. (eds.) 1988. *Latent Trait and Latent Class Analysis*, Plenum, New York.

Pedler, P. 1987. Accounting for psychometric dependence with a class of latent trait models. (Doctoral dissertation, University of Western Australia).

Rasch, G. 1968. A mathematical theory of objectivity and its consequence for model construction. European Meeting on Statistics, Econometrics and Management Science, Amsterdam.

Rasch, G. 1980. *Probabilistic Models for Some Intelligence and Attainment Tests*, 2nd edn. University of Chicago Press, Chicago, Illinois.

van der Linden, W. J. and Eggen, T. J. H. M. 1986. Am empirical Bayesian approach to item banking. *Appl. Psychol. Meas.* **10**, 345–54.

Weiss, D. (ed.) 1983. *New Horizons in Testing.* Academic Press, New York.

Wright, B. D. and Masters, G. N. 1982. *Rating Scale Analysis.* MESA Press, Chicago, Illinois.

Wright, B. D. and Stone, M. H. 1979. *Best Test Design.* MESA Press, Chicago, Illinois.

15 Individualized Testing in the Classroom

J. M. Linacre

Individualized testing can be quickly and easily instituted in the classroom by means of computer-adaptive testing. Computers in the classroom give teachers flexibility in the way that instruction is presented and tests are scored. Useful achievement measurements and performance diagnostic information can be obtained from even the first test administration with a small teacher-constructed item bank. The teacher can then continue to develop the item bank and administer subsequent related tests to the students. Each student's progress can be tracked and particular learning difficulties identified in a far more immediate and informative manner than with conventional paper-and-pencil testing. The algorithms underlying the test procedures can be based on Rasch measurement concepts and are simple to program and use. The data management requirements are not complex.

1 Computer-adaptive testing (CAT) redefined for the classroom

The theory required for the management of testing where different tests are administered to individual students based on their observed or expected performances is well-known (see *Adaptive Testing*). It was initially developed in the context of the availability of a large number of test items, a large number of students to test, and an expensive testing environment. The prime design consideration was that the tests administered be as short and efficient as possible. The goal was to obtain for each student one number representing that student's performance level.

In the classroom, however, the design considerations are quite different. There are fewer test items and fewer students, but detailed indications of students' strengths and weaknesses are often more important than overall performance level summaries. Timely reporting is essential, as information even a few days old may no longer be relevant to the classroom instruction currently under way. Compared to paper-and-pencil tests, individualized test administration by computer can streamline testing, reporting and record-keeping, and, after the initial set-up phase, lessen the clerical effort.

The theory generally applied to CAT is that each content area can be expressed in terms of an operationally defined variable representing a latent trait. Each item has a certain difficulty level and each student a certain ability in a content area. The probability of a student succeeding on a particular item can be modeled in a number of ways (see *Item Response Theory*). The Rasch model, however, not only has the most desirable measurement characteristics, but also the most satisfactory estimation characteristics for small samples.

2   Testing individuals in the classroom

The teacher begins with the area of curriculum or chapter in a text-book that is to be tested, and constructs test questions or "items" that follow from and probe this material. These items can be in multiple-choice format, but computer administration also facilitates free-form responses and automatic scoring of arithmetic problems and cloze questions.

To begin with, item writing can be time-consuming and require extra effort on the part of the teacher, but software aids, such as HyperCard ([TM]Apple Corp.), can lessen the burden. The art of writing items to probe the material that has just been taught is easily mastered and professionally satisfying. In CAT, useful information is obtained even when the initial set of teacher's items, now called an item bank, is only the test that would have been administered in a conventional classroom paper-and-pencil test. Items can then be added to or deleted from this bank as desired, clarifying the focus of the testing, and also developing and enlarging the teacher's testing resources.

Initially the specific difficulties of the items may not be well-known, but even stratifying the items by hypothesized difficulty is productive. The range of values of item difficulties is often set so that the middle difficulty items are given difficulty calibrations around 0 logits (log-odds units). Assigning initial calibrations of -1 logits (log-odds units) to the easy items, 0 logits to the medium items and $+1$ logits to the hard items is helpful both from the viewpoint of selecting items suitable for each examinee and for diagnosing unexpected behavior (Yao, 1991).

To administer these items to a student, a computer program is needed. Though there are elaborate published programs, for example, MicroCAT (Vale and Weiss, 1987), CALL (Eurocentres Learning Service, 1991) and intricate descriptions of professionally developed systems (Reckase, 1974; Nitko and Hsu, 1984), the actual computer programming required is simple, straight-forward and within the capabilities of any school's computer science department. Locally developed software is easier to tailor to local needs, and often much less expensive than professionally developed systems.

Administration of each test begins with the assignment of any reasonable initial ability estimate to the student. A starting ability estimate can be in the lower central part of the range of item difficulties, for example 0 to -1 logit, or else can be derived from the student's previous performances, for example, previously able students could start at $+1$ logit.

The CAT program then selects an item to present to the student. For statistical efficiency, this is of about the same difficulty as the student is thought to be able to deal with. There are situations, however, when it is desirable to give the student a feeling of having "succeeded" on the test in order to lessen test anxiety and other negative feelings (Gershon, 1992). Accordingly, the CAT program can deliberately select items that are slightly easy for the particular student. This does not affect the validity of the test, but can mean that a few more items must be administered in order to obtain a given level of measurement precision.

As each further item is administered, a better estimate of the student's ability is obtained. If the student succeeds on the first item, the ability estimate is increased slightly. Failure decreases the ability estimate slightly. An item appropriate to the revised ability estimate is selected and administered and the process repeated. As the number of items administered increases, the precision with which the ability is estimated also increases so that the standard error of the ability estimate decreases. The test can end when the standard error becomes less than some preset amount, or when the student has been administered the maximum number of questions considered relevant, or when there are no more relevant questions in the item bank, or when a time-limit is reached, or just when the student becomes tired.

As the CAT system develops in use, refinements are easily added. Further item selection and stopping rules can be introduced to investigate unusual response patterns and to ensure balanced use of the items in the bank. Students can be allowed to review and change previous answers or even to skip questions (Lunz et al., 1992). In particular, control over testing by the teacher permits variations in item format and presentation that are not feasible with standard CAT software or published tests.

Unless there is some test-coverage requirement which forces the administration of very easy or very hard questions, administering items on which there is less than 10 percent chance of failure or success, that is, more than 2.2 logits distant from the student's estimated ability, tends to provoke guessing or carelessness. Such off-target items do not expedite the measurement process. Indeed, it is the elimination of such inappropriate questions which is one of the main aims of CAT. Targeting item selection not only shortens the test but also reduces the provocation of irregular behavior and of the negative feelings so often associated with a test experience.

On completion of the test, the student can be informed immediately of the level of success by reporting an ability estimate rescaled from logits into some more familiar range of numbers, and also whatever diagnostic feedback the teacher deems useful. Reported to the teacher is the ability estimate and diagnostic information concerning which easier items the student unexpectedly failed and which harder items the student unexpectedly answered correctly. This gives the teacher a performance profile of each student's strengths and weaknesses relative to that student's overall performance level, thus enabling the teacher to give special, appropriate instruction to each student and to groups of students with similar profiles. Using graphical aids in these reports permits teachers with little numerical background to make immediate and accurate use of this information (Masters *et al.*, 1990).

As the items are administered, better information accumulates about their difficulty and validity. Poorly specified items can be revised or deleted. Item difficulties can be recalibrated to correspond to how difficult the students actually found the items. Then the next time these items are used, the measures they produce will have greater validity. New items can be added to the bank corresponding to each newly taught content area. The next test administered by the CAT program can thus include both new items and older items. This enables the relative difficulty of new and previous items to be compared, and, even more usefully, the progress of each student to be tracked in terms of increasing logit ability estimates. The use of previous items also reinforces student recall of earlier material.

The CAT item selection algorithm is used to prevent the administration of unproductively easy or hard items. Since the item bank can be entirely under the teacher's control, there is never the need to use an undesired item just because it happened to be in a published test or was used last year.

3 Adaptive-testing algorithms

The Rasch algorithms used in classroom CAT have as good or better theoretical measurement properties than the large-scale adaptive testing algorithms. They are simple to implement as computer programs on classroom-level computer equipment.

Estimation of the student measures can be carried out by applying maximum likelihood methods directly (Wright and Stone, 1979, pp. 15–20). A CAT program using such a method requires only a few pages of BASIC computer code. For classroom use, however, results which are identical for all practical purposes can be produced by even simpler approximations (Wright, 1988).

3.1 Testing to determine ability

Using the PROX algorithm (Cohen, 1979), a useful current estimate of student ability at any point during the testing process is given by accumulating the following information:

(a) The number of items administered so far = L.
(b) The number of right answers = R
     (if there are no right answers yet, then let R = 0.5).
(c) The number of wrong answers = W
     (if there are no wrong answers yet, then let W = 0.5).
(d) The sum of the item difficulties administered = SUMCAL.
(e) The sum of squares of those item difficulties = SUMSQCAL.
(f) Calculate the mean item difficulty administered, MEANCAL = SUMCAL/L.
(g) A dispersion factor for the spread of the item difficulties is
     $DF = \sqrt{(1 + (SUMSQCAL/L - MEANCAL*MEANCAL)/2.89)}$.

(h) The ability estimate, B, in logits is
     B = MEANCAL + DF*LOG(R/W).
(i) The precision of estimate, B, is the standard error of measurement, SE,
     SE = DF*$\sqrt{}$(L/(R*W)).
(j) The student's ability is estimated to be in the range B ± SE.
(k) Stop test, if stopping criteria are satisfied.
(l) Select the next item to administer randomly from the difficulty range, B ± 1, or, for
     an easier test, (B − 1) ± 1.

The range of items, from which the best next one to administer is selected at random,
can be defined by a range or window of item difficulties. Random selection of the next
item administered from a relevant range of items promotes balanced use of the items and
ensures that students of similar difficulties experience different tests, so reducing the
possibility of earlier test-takers coaching later ones. If the target success rate by each
student on the items is to be 50 percent, then the next item to be administered could have
a difficulty level within about ± 1 logit of B.
    When it is desired to give the student a positive experience in an examination so that the
student leaves the test with a general feeling of success, suitable, slightly easy items are
those on which the student has an overall 70 percent (say) chance of success. Such items
are those in the difficulty range centered about 1 logit less than B, the student's estimated
ability. The item selection window is the range of difficulties B-2 to B, that is, (B − 1) ± 1
logits.

3.2 Speeding up item presentation

Though computers can generally present items far faster than students can respond, very
able students may become frustrated if their almost instantaneous responses to simple
recall items cause them to wait for the computer to display each subsequent item.
    Speedier item presentation occurs when the computer selects the next item while the
student is still answering the current one. Since the difficulty of the selected items only
changes slowly, this has little effect on the statistical efficiency of the test.
    To obtain speedier item administration:

(a) Select two items initially.
(b) Display one of the selected items on the screen.
(c) Capture the student's response.
(d) Immediately display another selected item.
(e) While the student reads the displayed item, perform steps (a)–(l) of 3.1, based on
     the earlier item and the student's response in (c). Step (l) of 3.1 selects the next
     item.
(f) Loop back to (c), unless the test is to stop.

(g) If test is to stop, capture student's response to last displayed item, and calculate student's final ability estimate from steps (a)–(j) of 3.1.

3.3 Test anxiety

For various reasons, students may exhibit uncharacteristically low performance levels on the first few items of a CAT test. The computer can make note of the ability level estimated after, say, the first four items. If later ability estimates are significantly higher, then test anxiety or other factors may have degraded initial performance. A simple correction is to drop the first four items from ability estimation, recomputing the student's ability as though the test started at the fifth item.

3.4 Testing to determine mastery

When the purpose of the test is not to estimate a student's ability, but simply to decide whether that ability is above or below a certain criterion level, (see Criterion referenced measurement) a simple adjustment to the CAT algorithm suffices.

Start by administering an item close to the criterion level, C. Continue by administering items corresponding to the student's current ability estimate, using steps (a)–(l) of 3.1. Stop testing when the ability estimate is significantly distant from the criterion level, C. This occurs when the ability estimate, B, is at least one standard error (or, at most, two) from C. For a student with ability 1 logit from the criterion level, this would require a test of 15–20 questions.

There will be occasions when a student's ability level is too close to the criterion level to reach a pass/fail decision promptly. The test stops after some maximum number of items is administered. The decision relative to the criterion then depends on the purpose of the test. The student has neither clearly exceeded nor clearly fallen short of the criterion level. If the purpose is to verify mastery, then the student must clearly exceed the criterion level. If the purpose is to verify failure, then the student must clearly fall short of the criterion level.

3.5 Using CAT tests for diagnosis

From the teacher's viewpoint, diagnostic information about a student's performance is usually more valuable than a single, holistic measure of ability. Several types of information can be obtained from one test.

First, the content area can be divided into different strands or objectives. The student's ability on each strand can be estimated independently and thus a profile of the student's performance obtained which identifies strengths and weaknesses among strands. To facilitate this, the item administration process can be designed to select items from all strands in an even pattern.

Second, deficiencies or special strengths in the student's performance profile can be detected. In the course of the test administration, each student is likely to meet, or can be arranged to meet, a few questions that are somewhat hard and a few that are somewhat easy. If any of these off-target items provoke unexpected responses, the misfitting responses can be investigated for their diagnostic implications. Excessively hard items however, provoke wild guessing. Excessively easy items provoke specious answers to student-perceived "trick" questions.

Third, test items can be designed such that even incorrect answers provide useful diagnostic information about the student's performance (Adams, 1988). Partially correct answers can also focus item selection and assist in estimating student ability.

3.6   Calibrating the item bank

Item banks developed and maintained in the classroom have the advantage that test items can be added or deleted at any time. All estimates of item difficulty are, of course, only approximate. Even well-established items require periodic checks to ensure they continue to perform as intended. The deliberate inclusion, among the items used in each test session, of items in the same content area from other test sessions enables the entire item bank to be regarded as one comprehensive test, of which any particular student is administered a small section at any one time. Recalibration of the item bank enables the relative difficulties of all items in a content area to be obtained and maintained.

In order to validate or recalibrate the item difficulties, the responses made to the items by each student, along with that student's final ability estimates, are saved in an archive file after the test is completed. These archive files form a database used to obtain revised item difficulty estimates via standard Rasch estimation algorithms and computer programs or by using an algorithm equivalent to the one used for obtaining ability estimates.

Thus, for each item:

(a) The number of students who were administered the item $= N$.
(b) The number of right answers by them to the item $= R$ (if there are no right answers yet, then $R = 0.5$).
(c) The number of wrong answers by them to the item $= W$ (if there are no wrong answers yet, then $W = 0.5$).
(d) The sum of abilities of those answering the item $=$ SUMMES.
(e) The sum of squares of those student abilities $=$ SUMSQMES.
(f) Calculate the mean student ability for the item, MEANMES $=$ SUMMES/N.
(g) A dispersion factor for the spread of the student abilities is
     $DF = \sqrt{(1 + (SUMSQMES/N - MEANMES*MEANMES)/2.89)}$.
(h) The item difficulty estimate, D, in logits is $D = MEANMES - DF*LOG(R/W)$.
(i) The precision of estimate, D, is the standard error of measurement, SE,
     $SE = DF*\sqrt{(N/(R*W))}$.

Diagnostic information equivalent to that for the students can also be obtained for items. This indicates whether each item is functioning as intended.

4   Item bank and the curriculum

The item bank contains items written to represent a content area. After bank recalibration, the difficulty estimate of each item is an indication of how difficult the students found that item. Consequently, the difficulty ordering of the items is an empirical representation of the acquisition ordering of the content area (Wright and Bell, 1984). In general, a teacher presents the easiest material first, advancing to more complex material. The empirical ordering of items in the bank informs the teacher as to whether the teacher's organization of the material coincides with the student's experience. This can give the teacher greater confidence in the presentation sequence in use, or lead the teacher to alter the degree of emphasis or sequence of presentation of the material in order to expedite student learning.

5   Tracking student progress

Because the difficulty of each test is positioned in the metric frame of reference of the curriculum content area item bank, the ability of each student is also positioned within that frame of reference. During each subsequent testing session a further ability estimate for that student is obtained on the same metric. These ability estimates are directly comparable with items having numerically equivalent difficulty calibrations. Thus these items give a criterion meaning to that ability measure.

Over time, quantitative ability estimates for each student describe a developmental sequence and can be used to track student progress. This progress can be reported not only in terms of position relative to the rest of the class, but more substantively in terms of the particular items in the bank which have been mastered, and consequently in terms of the material in the curriculum which corresponds to the student's ability at any given time. The need for and type of special instruction thus becomes clear, and, when undertaken, its effect can be clearly observed by tracking the progress of the student's further ability estimates (see also: *Computerized Educational Testing*).

References

Adams, R. J. 1988. Applying the partial credit model to educational diagnosis. *Applied Measurement in Education* **1**(4), 347–61.

Cohen, L. 1979. Approximate expressions for parameter estimates in the Rasch model. *Br. J. Math. S.* **32**(1), 113–20.

Eurocentres Learning Service. 1991. CALL Computer Assisted Language Learning Authoring Program. Author, Zurich, Switzerland.

Gershon, R. C. 1992. Test Anxiety and Item Order: New Concerns for Item Response Theory. In: Wilson, M. (ed.) *Objective Measurement: Theory into Practice*. Vol. 1. Ablex, Norwood, New Jersey.

Lunz, M. E., Bergstrom, B. A. and Wright, B. D. 1992. The effect of review on student ability and test efficiency for computerized adaptive tests. *Appl. Psychol. Meas.* **16**(1), 33–40.

Masters, G. *et al.* 1990. *Profiles of Learning: The Basic Skills Testing Program in NSW*. Australian Council for Educational Research, Hawthorn, Victoria, Australia.

Nitko, A. J. and Hsu, T.-C. 1984. A comprehensive microcomputer system for classroom testing. *J. Educ. Meas.* **21**(4), 377–90.

Reckase, M. D. 1974. An interactive computer program for tailored testing based on the one-parameter logistic model. *Behavior Research Methods and Instrumentation* **6**(2), 208–12.

Vale, C. D. and Weiss, D. J. 1987. *MicroCAT Testing System*. Assessment Systems Corporation, St. Paul, Minnesota.

Wright, B. D. 1988. Practical adaptive testing. *Rasch Measurement* **2**(2), 21.

Wright, B. D. and Stone, M. H. 1979. *Best Test Design*. MESA Press, Chicago, Illinois.

Wright, B. D. and Bell, S. R. 1984. Item banks: What, why, how. *J. Educ. Meas.* **21**(4), 331–45.

Yao, T. 1991. CAT with a poorly calibrated item bank. *Rasch Measurement* **5**(2), 141.

Further reading

Alderson, J. C. and North, B. (eds.) 1991. *Language Testing in the 1990s: The Communicative Legacy*. MacMillan, London.

Baker, F. B. 1986. Item banking in computer-based instructional systems. *Appl. Psychol. Meas.* **10**(4), 405–14.

Schoonman, W. 1989. *An Applied Study on Computerized Adaptive Testing*. Swets and Zeitlinger, Rockland, Massachusetts.

16 Item Analysis in Test Construction

J. J. Barnard

1 Introduction

Testing seems to be a very straightforward process. Simply compile a test by asking a number of hopefully relevant questions, administer it to a group of people, score the answers according to some scheme and express the results on a scale. In truth it is, however, not quite so simple to locate individuals on a continuum with respect to a particular psychological construct. Human attributes are complex, difficult to define and cannot be measured directly. Measurement techniques that are able to yield valid, reliable, objective and efficient measures of latent traits are therefore not trivial.

Tests represent one particular measurement technique designed for use as systematic procedures for studying the behavior of an individual or a group of individuals (Cronbach, 1970; Ebel and Frisbie, 1991, p. 26). In essence, a test is a series of tasks or a set of questions (items) that persons answer orally or in writing which makes it possible to determine defined differences between persons.

This chapter describes and contrasts two analytic procedures that are commonly used in test development, namely traditional or standard item analysis within the framework of classical test theory (CTT) and modern analysis which has its roots in item response theory (IRT). These processes usually follow the identification of the purpose for testing and preparation of a pool of items in the process of test construction.

2 The purpose of testing

In general, the many and varied purposes for testing can be grouped into two main categories, namely a bureaucratic category and a professional category. The former mode refers to control, monitoring and certification and aims at fulfilling the functions of summative assessment. The latter mode is connected with student learning enabling the teacher to determine whether growth in knowledge and understanding has taken place; whether effective learning and teaching have been accomplished; and whether realistic standards of achievement are maintained. As good tests very seldom serve multiple purposes equally well, it is necessary to decide how the test scores will be used. The interpretation of scores determines the point of reference which will be used to obtain meaning from the scores, whether a person's optimal performance or typical performance is of importance. The uses of the test and the people for whom the test is intended play an important role in the construction of the test model. They determine the properties of the test such as test length.

3   Prepare a pool of items

Once the purpose for testing has been determined, the construct to be measured needs to be defined as clearly and completely as possible. The test constructor can then delimit the field of interest, assure that the facets of the construct rest upon a sound theoretical basis, consider the relative weights allocated to each aspect in the construct as it related to the content validity and the rationale of the test and decide how the answers to the questions will be scored. Decisions on these and related issues, yield the test model which provides guidelines according to which the items of the test are written.

Typically the test constructor will conceptualise one or more types of behavior which are believed to manifest the construct and then write items that require these behaviors to be demonstrated. To Ebel and Frisbie (1991, pp. 117–8) the procedures enumerated for addressing intrinsic rational validity include a description of the abilities of interest and domain of knowledge and skills to be tapped, decision on the relative importance of the various subdomains, and a decision on which types of items will best require the demonstration of relevant knowledge. This test plan will guide the test constructor to determine the type(s) of items to be used, the number of items of each type, the kinds of tasks the items will present, the number of tasks of each kind needed, descriptions of content areas to be sampled, the number of items for each area needed, and the level and distribution of the difficulty of the items.

For norm-referenced interpretation where an individual's performance is compared with the performance of others, the content domain can be defined more generically, whilst an explicit description of the content specifications is needed for a domain-referenced interpretation of scores to emphasise diagnostic assessment and to some extent formative assessment. In criterion-referenced testing the test is primarily intended to certify whether an individual has attained a certain level of minimum competency and therefore a series of sets of items in which each set corresponds to a determined level of proficiency on the criterion is included in the test.

Test constructors often outline the coverage of substantive content and the cognitive processes of a test by means of a table of specifications which shows the relative emphasis that each component receives in the test. The areas of content are usually assigned to the rows of a two-way grid and the kinds of the abilities, for example, those given in Bloom's taxonomy, to be assessed are assigned to the columns of the grid. All the items in the test are classified in one of the cells and the test constructor can decide the proportion of items from each cell (see Bloom *et al.*, 1971). Although the number of items in each cell usually indicates the relative emphasis or importance in the test, the score points represented by each cell should rather be considered when items are not scored dichotomously.

Once the items have been written and classified, they should be given to a panel of experts to review. Their comments as well as a try-out with a small number of persons from the population for which the test is intended is very useful in monitoring the quality of the items and in finding language problems and technical flaws in items.

The items are now ready to be field tested on an appropriate sample of persons. According to Nunnally (1978, p. 11), psychometric theory is a large-sample theory because large numbers of persons are usually required for test development and validation procedures. Sample sizes may vary according to the item analysis procedures and need for cross validation or not. Should only classical indices be required, Nunnally's rule-of-thumb to have five to ten times as many persons as items should suffice (Nunnally, 1967). If item response theory is used as the underlying test theory, samples between 200 and 1000 persons will be needed, depending on the particular model employed.

At this stage at least one and a half times as many items as required in the final test should be field tested to enable the test constructor to select items for the final test. The aim of this administration is therefore to collect information which can be used to select the items for the final test with the required characteristics, to determine the length of the test, detect possible bias in items, and to evaluate the difficulty, reliability and validity of the test.

4   Traditional item analysis and selection

Item analysis is not a substitute for the originality, effort and skill of the item writer and relatively poor statistical results can be overruled on logical grounds. Although an examination of the statistical properties of item scores provides the test constructor with useful and necessary information, item analysis often only identifies problems, and the test constructor must search for the probable causes and possible solutions (Ebel and Frisbie, 1991, p. 225).

Item analysis begins after the items have been scored. Items are scored either dichotomously as is often the case for multiple-choice items or polychotomously as is often the case for short-answer items or essays. Information about the score distribution of each item and its relationship to the other items is obtained because items are the building blocks of tests. A test can have no property that is not a function of the items that comprise it (Crocker and Algina, 1986, p. 87).

The main aim of item analysis is to obtain objective information about the items which can be used to indicate defects in items and identify poor items. Items may be too easy or difficult or fail to discriminate adequately between high and low achievers.

The two most basic statistics computed and examined during item analysis are the items' difficulty and discrimination values.

4.1 Item difficulty

Item difficulty (p) is defined as

$$p_i = \frac{1}{N_t} \sum_{j=1}^{N_t} X_{ij}$$

where

p_i is the difficulty value of item i

N_t is the number of persons in the sample

and

X_{ij} is the item score on item i for person j.

If the item is dichotomously scored, the difficulty value of the item is equal to the proportion of persons who answered the item correctly relative to all the persons tested, i.e. $p_i = \frac{N_c}{N_t}$ where N_c is the number of persons who answered the item correctly. The higher the value of p_i, the easier the item. This definition can be amended by letting N_t be equal to only the number of persons who responded to the item. This change can become significant especially for items at the end of a test where more missing values are usually found. In practice a distinction between items that were omitted and items that were not reached is made during the item analysis stage.

4.2 Item discrimination

Optimal performance tests such as achievement tests purport to provide information about individual differences, i.e. distinguish between individuals who are high and those who are low on the criterion of interest. Mostly the total score is the only adequate measure available of the construct and is thus used as an internal criterion to identify items for which high-scoring individuals have a high probability of answering correctly and low-scoring individuals have a low probability of answering correctly. Ebel and Frisbie (1991, p. 227) point out that an external criterion has no advantage over an internal criterion unless it is truly a better indicator of whatever the test is supposed to measure.

Different statistics can be computed to describe the degree of relationship between responses to individual items and the total test score. It is common practice for items that are scored dichotomously to determine the correlation coefficients such as biserials and points biserials which describe the relationship between the item score and the total test score for each person, phi coefficients when scores from a dichotomously scored item are correlated with scores from a dichotomous criterion, and tetrachoric correlations for correlations between two dichotomous variables. According to Crocker and Algina (1986,

p. 313), which coefficient is used depends on various factors such as the difficulties of the items and the abilities of the individuals. However, the point biserial correlation is used mostly in practice because it does not involve restrictive assumptions and is therefore more generally applicable. It differs computationally and theoretically from the biserial coefficient, and yields values that are always at least a quarter less than the biserials, but the two can be interpreted in essentially the same manner (Ebel and Frisbie, 1991, p. 232). If items are not scored dichotomously, the Pearson product moment correlation can be computed in correlating item scores with criterion scores.

For dichotomously scored items the point biserial correlation is defined as

$$r_{it} = \frac{\mu_c - \mu_t}{s_t} \sqrt{\frac{p}{q}}$$

Where

> r_{it} is the correlation between the item and the total score
> μ_c is the mean score of the persons who answered the item correctly
> μ_t is the mean score of all the persons
> s_t is the standard deviation of the scores of the group
> p is the proportion of the persons who answered the item correctly

and

$$q = 1 - p$$

This coefficient is a particular case of the product moment correlation coefficient.

Since the item score contributes to the total score of each individual, the correlation between the item and total score will be inflated and a correction is necessary if the index is computed for a small number of items. An adjustment such as the one proposed by Guilford (1966) which yields a lower value may be used in such cases.

This statistic thus refers to the extent in which a test succeeds in distinguishing between persons who performed well in the test in general and persons who performed poorly in the test. The higher the value the better the item complies with this purpose and in that sense the discrimination value gives an indication of the quality of the item.

Two indices that are both a function of item variance and item score correlation with a criterion which are often computed during item analysis are the item reliability index and item validity index. The former, also sometimes referred to as Gulliksen's product, is defined as the product of the item's discrimination value, r_{it} and the standard deviation of the item for dichotomously scored items. If an external criterion is used, the item validity index is defined as the product of the item-criterion correlation, r_{ic} and the standard deviation of the item. These two indices may be of value in some test construction stituations and provide useful information about the test as a whole after final selection of items (Crocker and Algina, 1986, p. 320).

5 Traditional test construction

Once item statistics for the pool of items have been computed or estimated, the test constructor can select appropriate items for the test according to their content and statistics. Through the ongoing process of item selection, the test constructor should decide what properties the test should have and remain focused on the purpose of the test. Ebel and Frisbie (1991, p. 221) name relevance, balance, efficiency, specificity, difficulty, discrimination, variability and reliability as desired properties of a good test.

Perhaps the most important characteristic of a test is its validity, especially since test scores are usually used to draw inferences beyond the testing situation. It should be pointed out that a test is not valid or invalid in itself, but is valid to the purpose for which it is used. Any test should measure what it purports to measure. Different types of test place different emphases on different types of validity. Where construct and criterion-related validity are perhaps more important for psychological tests, content validity is most important in tests such as achievement tests.

Good tests are reliable, i.e. yield consistent results over repeated administrations. Where (content) validity is more a qualitative decision, the reliability of a test can be expressed as a number. Traditionally test reliability refers to the relationship between true scores, observed scores and errors of measurement and is defined as the ratio of true score variance to observed score variance (Guilford and Fruchter, 1978, p. 410). But, since the relative contributions of true score and error variance to observed score variance are unknown, reliability cannot be estimated before further steps are taken – usually by defining a second measure and assuming that persons have the same true scores as on the first measure, but with independent errors of measurement and equal variances of the errors of measurement for the two measures.

By applying the same test twice, a test-retest reliability coefficient can be computed. It is however, not possible to guarantee the required stability of true score and randomness of measurement errors. In order to avoid the problem of possible changes in true score, the parallel forms method or split-halves method can be applied. The reliability estimate obtained for the half test can be improved to indicate the reliability of the whole test by using a special version of the Spearman-Brown formula. But, whether parallel forms can be achieved in practice is questionable (Hambleton and Swaminathan, 1985, p. 2).

While the test-retest and parallel forms methods are ways of assessing the stability of measurement, the split-halves method yields an estimate of the homogeneity or internal consistency of a test. Because of a dissatisfaction with the arbitrary splitting of tests into halves which can yield halves which are more or less equivalent than others, several other methods have been developed. Cronbach's coefficient alpha (Cronbach, 1951) and Kuder-Richardson 20 formula (Kuder and Richardson, 1937) are perhaps the best known.

Closely related to reliability is the standard error of measurement which may be considered as the average standard division of the error distributions of persons for a large number of repeated testings. The standard error of measurement is reported as a single, global statistic for the whole group.

It should be noted that the reliability of a test is determined by factors such as the heterogeneity of the test group, the abilities of the persons, the variation in item difficulty, the technique used to determine the index and length of the test. Both reliability and validity are, to some extent, dependent on the variance of scores. By adding more items to a test the total test variance will increase by the sum of the items' variances and their covariances with all other items in the test. It should be noted that the sum of the variances for individual items is less than the variance of the total test scores since the total variance is determined jointly by the variances of the individual items and by the covariances of all pairs of items. In fact the total variance for a test with x items will have x variance terms and $x(x-1)$ covariance terms of which $x(x-1)/2$ are unique.

There is general consensus among psychometricians that a test of minimum length that will yield scores with the necessary degree of reliability and validity for the intended uses is desirable. Since the items in a test are a sample of all possible questions that can be asked, the items should be representative of the domain to the extent of the thoroughness to which the domain must be sampled. Although increasing the number of items will generally yield a smaller sampling error and higher reliability coefficient, the amount of testing time is not unlimited and longer tests introduce other factors such as fatigue.

How difficult should a test be? As Ebel and Frisbie (1991, p. 223) rightly point out, if difficulty were strictly a characteristic of the test, a given test would be equally difficult or easy for every group to whom it was administered. The difficulty level of a test relates to the purpose for testing and the kind of score interpretation desired. In, for example, achievement tests the purpose for testing is to separate individuals and therefore large standard deviations of scores are desirable, i.e. test variance should be maximised. Too easy and too difficult tests will both have skewed distributions with relatively small standard deviations. For this purpose, tests comprising items with moderate difficulties will produce the highest score variability.

For dichotomously scored items, the variance of an item is given by pq and therefore the ideal difficulty value seems to be 0.5 as this value yields the highest possible item variance, namely 0.25. If multiple-choice items are used, guessing tends to inflate p-values and therefore items with difficulty values slightly higher than 0.5 seem to give the best discrimination. In practice items with difficulty values at a point halfway between perfect and the chance level difficulty seem to give the best results. It is obvious that if a test for a scholarship is to be constructed, more difficult items should be included in the test.

Some educators believe that some easy items should be placed at the beginning of a test and some difficult items at the end of a test. From an educational point of view this strategy might put the persons responding at ease and challenge the more able ones, but from a statistical point of view the easy items merely add a constant to each person's score without affecting the rank order of scores in general.

After the items for the test have been selected and combined as a test, estimates of the statistical properties of the test can be made. The mean of the test can be estimated by adding the difficulty values of the items included in the test. High stakes tests are, after compilation, often administered to an independent sample for cross validation.

6   Methods using item response theory

Classical test theory has been the backbone of educational and psychological measurement for most of this century. While syntactically correct, there are many well-documented shortcomings which give rise to some serious practical problems associated with the model (Hambleton and Swaminathan, 1985). It provides no test-free scores, has sample-dependent statistics, leans heavily on the availability of parallel measurements and provides no basis for determining what an individual might do when confronted with an item.

By definition, item difficulty depends on the general ability of the person responding, item discrimination as well as reliability depend directly on the heterogeneity and distribution of the abilities of the persons, and abilities of persons are interpreted in terms of the number-right score which is dependent on the specific test and its difficulty. Even if difficulty values are converted to standard scores (see Guilford and Fruchter, 1978, p. 458) there will be systematic differences depending on the levels and ranges of the groups on whose responses the computations are based. Furthermore, biserial correlations for ostensibly equivalent forms have shown considerable variation in practice, beyond that which might be expected from sampling theory (Wood, 1976, p. 255).

The usefulness of a test compiled by classical techniques is therefore determined by the resemblance of a test group for which the test is intended and the group of persons used to determine the item indices (Hambleton, Swaminathan and Rogers, 1991, p. 99). This problem has been recognised for many years (Gulliksen, 1950, p. 392; Lord and Novick, 1968, p. 328). Furthermore, items cannot be selected to yield a test that meets a certain specification in terms of measurement precision since an item's contribution to the reliability of a test does not depend on the item only, but also on the relationship between the item and the other items in the test.

The need has been recognised for scales other than those provided by simple counts of correct responses, especially when the tests are of different difficulty or have different numbers of items. Item response theory (IRT) purports to overcome these problems through offering person-free measures of item characteristics and item-free measures of person abilities (Crocker and Algina, 1986; Hambleton and Swaminathan, 1985). The essence of IRT, namely that item difficulty and person ability are measured on the same scale, makes it possible to select items to function optimally in predetermined ability regions.

IRT is a statistical theory consisting of mathematical models expressing the probability of a particular response to an item as a function of the abilities of the persons and of certain characteristics of the item. The three-parameter logistic model has item parameters for difficulty and discrimination power as well as a pseudo-chance parameter, which is sometimes referred to as a guessing parameter. When the probability of a correct response to an item is expressed as a function of ability, this expression is referred to as the item characteristic curve (ICC). The difficulty of an item is located at the point of inflection of the ICC, i.e. where the probability of answering the item correctly is 0.5, if no guessing is possible. The discrimination parameter is associated with the steepness of the ICC and the pseudo chance parameter is associated with the probability that a person lacking the ability being measured is able to succeed on the item. The two-parameter model assumes that a person with low ability has no chance of succeeding on the item, not even by guessing. It should be pointed out that the difficulty and discrimination parameters are independent of each other although the item's difficulty is defined as the point at which the slope of the ICC is at a maximum. The one-parameter (or Rasch) model assumes that all items discriminate approximately equally and allocate a constant value to the discrimination parameter.

Estimates for the ability of each person and for the item parameters of the chosen model are obtained from the response data. Various iterative procedures for the joint calibration process have been developed and refined. Maximum likelihood estimation is preferred by many practitioners because it represents a conventional statistical approach and has desirable and useful properties such as consistency and efficiency provided that there are no persons with zero or perfect scores or items which have been answered either correctly or incorrectly by all persons, while Bayesian methods have also received increased attention.

Once parameter estimates have been obtained for the model chosen, the suitability of the item response function can be assessed. Approaches for assessing goodness of fit include the checking of model assumptions, checking expected model features and checking model predictions of actual and simulated test results. An evaluation of how well the model chosen fits the data is essential in IRT since it provides the test developer with information on the consistency of responses at the item level as well as about the quality of the analysis and hence the usability of the results. In general, it can be said that if only a few items misfit, they may be poor items, but if many items do not fit, a more general model should be fitted to the data. No model will, however, fit poor items.

Methods of detecting poor items in IRT are not as straightforward as in CTT where the quality of an item can be judged on the basis of its difficulty value and discrimination value computed from relatively small samples of persons. In the classical context an item analysis usually also includes a frequency count (or proportion endorsing) of the different options which provides useful information about the item.

There are currently many different procedures which are used in the examination of item fit to the Rasch scale. These different procedures are employed in the different computer programs that are used in the estimation of item, person and facet parameters.

These tests and other less widely used procedures that are employed to determine whether either dichotomous and polychotomous items fit the undimensionality requirements of the Rasch model have been classified by Wu (1997) into the following classes:

(1) chi-squared goodness of fit tests that are based on comparing observed and expected counts of various types;

(2) tests that compare standardised residuals to form approximately normal variates, that are based on comparing the observed and expected responses of individuals to items; and

(3) exploratory non-parametric tests that provide diagnostic information about specific model violations (Wu, 1997, p. 8).

Wu has investigated the derivation of a number of fit statistics for items, persons and raters and has presented both empirical evidence and theoretical justification to support the use of these indices in data analysis. The fit statistics that Wu has considered are the standardised residuals of the sufficient statistics for the parameters of the model. All persons are used without the arbitrary assigning of persons to fractiles thus avoiding the problems of low cell frequencies and the misclassification of persons to fractile groups. The sample distributions of the transformed weighted or infit t-statistics are found to be invariant with respect to test length and sample size and these statistics, in general, provide a stable frame of reference for comparisons and the testing of degree of fit to the Rasch model. This infit t-statistic is not greatly sensitive to unexpected observations, whereas the outfit or unweighted t-statistic occasionally produces outlying values. Consequently, the use of the infit or weighted statistics is to be preferred to the outfit or unweighted statistics. These fit statistics are found to have considerable strength with respect to certain violations of the requirement of unidimensionality, particularly lack of parallelism in the item characteristic curves. However, the infit or weighted t-statistic is not very sensitive to the occurrence of guessing. Moreover, in more complex models the fit values are influenced considerably by the constraints placed on other parameters in the estimation process. Since, the fit of items, persons and raters to the Rasch model is of great importance because of the requirement of unidimensionality, the strength and utility of the many different tests of fit need further consideration by research workers.

In order to obtain a measure of the precision with which a parameter has been estimated, the value of the standard error is usually computed. In contrast to CTT, the standard error varies along the scale which implies that abilities, for example, are estimated with various degrees of precision. Generally the most precise estimates of parameters are found when person abilities and item difficulties are closely matched.

Directly related to the standard error, is the concept of information which is defined as the reciprocal of the square of the standard error. By definition the amount of information provided by an item will vary along the scale. In addition to the invariance advantages over concepts such as reliability in CTT, by using information functions items can independently of each other be selected on the basis of the amount of information they provide to build a test item-by-item according to predetermined specifications (Wright and Stone, 1979; Hambleton and Swaminathan, 1985). Estimates of the amount of information a test provides, and the associated measurement precision, is provided at each ability level and is simply the sum of the information functions of the items in the test. Besides the obvious advantages this implies over CTT methods of compiling tests, the information needed for setting up norms for a standardised test, for example, could be obtained using a relatively small sample if IRT methods were used, provided that the persons are fairly well-matched with the items.

7   Conclusions

Even though item parameters are not, strictly speaking, defined in CTT, test developers find them useful to identify poor items and subsequently select items in order to compile a test. Some of the most important practical problems associated with the model have been emphasised. Lord (1980) stressed the necessity in test development to be able to predict the statistical and psychometric properties of a test when it is administered to any group, not only to one for whom a suitable standardising sample has previously been tested. The models of IRT make this possible as many studies that demonstrated the profound implications for the design and development of tests have shown. Where classical item statistics are mostly used in item selection only, IRT item parameters are also used in scoring, calibrating and equating tests.

In accordance with Hambleton, Swaminathan and Rogers (1991, p. 58) and Hulin, Drasgow and Parsons (1983, p. 67) it can be concluded that CTT and IRT should not be viewed as rival theoretical frameworks. A duet rather than a duel between CTT and IRT will provide most information to the test developer. The results obtained from a CTT based item analysis can yield useful information in finding flaws in items and guiding the test developer towards choosing an appropriate IRT model. The advantages that IRT parameters offer should subsequently be used for constructing tests for specific purposes, implementation of computerised adaptive testing, investigating bias, and the equating of different tests.

References

Bloom, B. S., Hastings, T. J. and Madaus, G. F. 1971. *Handbook on Formative and Summative Evaluation of Student Learning*. McGraw-Hill, New York.

Crocker, L. and Algina, J. 1986. *Introduction to Classical and Modern Test Theory*. Holt, Rinehart and Winston, New York.

Cronbach, L. J. 1951. Coefficient alpha and the internal structure of tests. *Psychometri.* **16**, 297–334.

Cronbach, L. J. 1970. *Essentials of Psychological Testing*. Harper and Row, New York.

Ebel, R. L. and Frisbie, D. A. 1991. *Essentials of Educational Measurement*. Prentice Hall, New Jersey.

Guilford, J. P. and Fruchter, B. 1978. *Fundamental Statistics in Psychology and Education* (6th Edition). McGraw-Hill, New York.

Guilford, J. P. 1966. *Psychometric Methods*. McGraw-Hill, New York.

Gulliksen, H. 1950. *Theory of Mental Tests*. Wiley, New York.

Hambleton, R. K., Swaminathan, H. and Rogers, H. J. (1991) *Fundamentals of Item Response Theory*. Sage Publications, Newbury Park, California.

Hambleton, R. K. and Swaminathan, H. 1985. *Item Response Theory: Principles and Applications*. Kluwer Nijhoff, Boston.

Hambleton, R. K. and Murray, L. N. 1983. Some goodness of fit investigations for item response models. In Hambleton, R. K. (ed.) *Applications of Item Response Theory*. Educational Research Institute of British Columbia, Vancouver.

Hulin, C. L., Dragow, F. and Parsons, C. K. 1983. *Item Response Theory*. Dow-Jones Irwin, Homewood, Illinois.

Lord, F. M. 1980. *Applications of Item Response Theory to Practical Testing Problems*. Lawrence Erlbaum, Hillsdale, New Jersey.

Lord, F. M. and Novick, M. R. 1968. *Statistical Theories of Mental Test Scores*. Addison-Wesley, Reading, Massachusetts.

Kuder, G. F. and Richardson, M. W. 1937. The theory of the estimation of test reliability. *Psychometri*. **2**, 151–160.

Nunnally, J. C. 1967. *Psychometric Theory*. McGraw-Hill, New York.

Nunnally, J. C. 1978. *Psychometric Theory*. (2nd edition). McGraw-Hill, New York.

Woods, R. 1976. Trait measurement and item banks. In De Gruijter, D. N. M. and van der Kamp, L. J. Th. (eds.) *Advances in Psychological and Education Measurement*. Wiley, London.

Wright, B. D. and Stone, M. H. 1979. *Best test design: Rasch measurement*. MESA Press, Chicago.

Wu, M. L. 1997. The development and application of a fit test for use with marginal maximum likelihood estimation and generalised item response models. (Unpublished Master of Education thesis) Melbourne: University of Melbourne.

17   Item Banking

J. Umar

Availability and quick access to good quality items is usually expected by both teachers and test developers. A large collection of good items will help teachers to concentrate more on their teaching without having to spend much time on item construction. It could also ensure that only high quality items are used. When such a collection (popularly referred to as an "item bank") consists of items measuring the same thing and calibrated onto a common scale, it could help test developers in solving many of the practical testing problems. Use of a calibrated item bank could thus affect policies in educational testing and assessment. This article discusses the concept of an item bank, its rationale, practices and the problems in its development and management.

1   What is an item bank?

There is no single agreement on how "item bank" is defined. It lies on a continuum from a very loose and unrestricted definition such as "any collection of test items" to "a relatively large collection of easily accessible test questions" (Millman, 1984), up to a quite restrictive definition such as in Choppin (1981a): "collections of test items organized and catalogued to take into account the content of each test item and also its measurement characteristics (difficulty, reliability, validity, etc.)" In fact, an item bank could be defined differently depending upon the purpose of its uses.

Despite the different levels of restrictedness in the definitions, there is one common ground for the definitions: only "good items" are to be stored in the bank. It is differences in the levels of what is meant by "good items" that makes the definitions vary. At the basic level, an item could be taken into the bank if it is constructed properly and its content is considered valid. See Popham (1978) or Roid and Haladyna (1982) for methods of item writing and validation. This type of item bank may be useful to teachers in preparing classroom assessment, especially when a total score or a scale is not important in the interpretation.

The next level of item banking is the inclusion of "traditional" empirical validation of the items as an additional criterion in item selection. At this level, items satisfying criteria at the basic level above are pilot tested, and item selection is made based on how well they behave as expected. Here, classical psychometric properties of items such as proportion, correct correlation between item and total scores (discriminating power), and distribution of responses to items distractors (in multiple choice items), are recorded. An item bank of this level could be useful to local test developers and school districts, as well as to teachers. Many examples of item banking of this type were reported in Brzezinski and Hiscox (1984). In the past, many major test publishers tried to develop this kind of item collection in an effort to make available a sufficient number of items when several parallel forms of a standardized test were to be constructed.

A higher level of item banking is a calibrated item bank. Here, items satisfying criteria of the basic level above are pilot tested in order to verify their fit to an item response model. The items which do fit the model are calibrated using a scale defined by the model. Calibration in this case involves defining the positions of individual items on a scale measuring both item difficulty and person ability. Based on this type of item bank, it is possible to design and construct a test which is expected to provide optimal information on the person's characteristic being measured, and with a high or even a desired degree of precision. Validation and calibration of items, test design and construction as well as test scoring under this level of item banking are made possible through the application of Item Response Theory (IRT). Hence, item banking of this type can not be separated from IRT itself. In fact, an item bank at this level could be considered as a model of a "measurement system". In this system, any new items intended for measuring the same attribute could be validated and calibrated onto the existing scale of the bank. Since the items are calibrated, it is possible to compare results from tests consisting of different subsets, of items from the bank (see *Item Response Theory*).

Another way of defining levels in item banking is by looking at how a bank is organized. As previously mentioned, one of the intended features of an item bank is that the items are "easily accessible". This could mean the involvement of computers. Items stored in a computerized data base should provide greater accessibility and efficiency. In the case of a calibrated item bank, it is nearly impossible to develop and operate such an item bank without a computer. According to Hambleton (1986), the failure of the first efforts in banking test items in the late 1960s and early 1970s in the United States and the United Kingdom was due to lack of computer software and facilities, because the amount of paper and administration was tremendous. Based on the extent of computer involvement in its operation, item banking could also be classified into: (a) fully manual item banking, (b) manual item banking using item cards but with computer services used in data analysis for item validation, and (c) a fully computerized item bank. The choice of level of computerization depends upon the purpose in banking test items, local conditions and situations.

It can be concluded from the above discussion that item banks can be either unrestricted and fully manual, or calibrated and fully computerized. However, since the types are hierarchical the discussion in this entry is concerned with the development, management and problems of the highest level of item banking, which is calibrated and fully computerized.

2   Why item bank?

The idea of item banking is associated with the need for making test construction easier, faster and more efficient. In the United States, the concept of item banking has also been connected with the movements to both individualized instruction and behavioral objectives in the 1960s (Hambleton, 1986). Van der Linden (1986) viewed item banking as a new practice in test development, as a product of the introduction of Item Response Theory (IRT), and the extensive use of computers in modern society. Therefore, when a large collection of good items is available to either teachers or test developers, much of the burden of test construction can be removed. The quality of tests used in the schools, for example, could be expected to be better than it could be without an item bank. When a calibrated item bank is developed under IRT, testing programs can be made more flexible and appropriate, because different groups of students can take different tests which are suitable to each of them and the results can still be compared on the same scale. Together with sophisticated computer software, application of computerized adaptive testing (Hambleton *et al.*, 1991) could be made possible at the school or district level. Other advantages of calibrated item banking include the following (Umar, 1990).

(a) The decentralization policy of a national testing program can be introduced without sacrificing comparability of results.

(b) Cost and time spent on test construction activities can be reduced dramatically.

(c) As the number of items in the bank becomes larger, the problem of security such as item leakage, becomes less important.

(d) Quality of testing programs, in general, can be improved because good items are easily available to users, especially teachers and local test developers.

(e) Teachers can design their own assessment instruments using relatively good items by sharing items in the bank.

(f) The best possible test for a given purpose or for a particular group of examinees can be designed.

(g) Since the basis for measurement is the item rather than the test, it facilitates a criterion-referenced interpretation.

(h) Teachers can concentrate their effort on teaching without having to spend much time on item construction.

(i) According to Choppin (1976), an item bank is suitable for countries with a large school system but with limited financial resources and psychometric expertise, because an item bank can provide a cheap but comprehensive system of educational assessment.

(j) Calibrated items can provide a systematic specification of what is important in the subject content. A curriculum can be visualized as a family of learning strands, and item calibrations (when the empirical ordering is valid) provide a curriculum map from which teaching strategies can be designed and against which rates of learning can be calculated.

(k) When an item bank is locally built, according to Millman (1984), the sense of ownership appears to be an advantage.

Choppin (1981b) also identified specific advantages of item banking for the development and operation of a system of national examinations. The advantages are classified into four categories.

(1) *Economy.* Thousands of high quality items are written by some of the finest teachers and examiners in almost every country in the world every year, and are used only once and then discarded. Under an item banking system, repeated use of items is possible.

(2) *Flexibility.* A calibrated item bank offers the facility for tailoring tests to specific applications. The tests may be long or short. Test coverage may be for a wide range of ability or focused at a particular level; covering the entire curriculum of focusing on a narrowly defined area. For any test constructed using calibrated items in a bank, individual items can be removed or added with predictable effects on the test's characteristics. Even a fully adaptive test in which every student can be exposed to a different set of items is potentially possible, and yet the comparability of test scores still holds.

(3) *Consistency.* In an item bank where all items are calibrated onto a common scale, the measurement system has a high degree of coherence and consistency which is not obtainable from networks of standardized tests. It is possible to construct parallel tests with equal meaning for the same score regardless of with which test form the score is associated. It is also possible to construct nonparallel tests such as an easier version for schools with less advanced students.

(4) *Security.* In many countries, examinations such as national school leaving examinations and university entrance examinations play very important roles. Results from such examinations may affect an individual's future. In this situation, pirated versions of test items or leakage of test content might pose serious problems. To illustrate how the security aspect of an examination is so important in some countries, it is possible to find a country where the national examination papers are printed abroad and heavily guarded (involving the police department) until the day of the examination. There are two ways in which item banking could relax security tension. First, as the number of items becomes larger, there will be thousands of items with comprehensive coverage of the entire curricular domain stored in the bank. Of course, it is no easy matter to learn the answers to such thousands of items without understanding the background material. Hence, even when the text of all items in the bank is published, the security of the examination would no longer be a serious problem. Second, it is quite easy to construct several

alternative forms of a test (without any loss of comparability of the scores) from a calibrated item bank, thereby alleviating the security problem. The particular form to be administered to a particular individual could be kept confidential until the final moment of the examination. In case of leakage of test items to one group of examinees, it would be relatively easy to substitute the test.

A number of educators and researchers have expressed reservations about item banking. For example, Baker (1986) argued that test development using item banks may not be as simple as it is claimed. Goldstein (1981) offers a number of reservations and criticisms when a calibrated item bank is built under the Rasch model. Some technical problems encountered in the establishment and operation of item banking are often considered as weaknesses or disadvantages of item banking. However, all of the reservations, criticisms and disadvantages are problems to be overcome before proper and successful item banking can take place, rather than a rejection of the idea of item banking. Such problems are discussed in a later section of this article.

3   Establishment of a calibrated item bank

There are several important activities involved in the development of a calibrated item bank, namely: (a) item writing, (b) item validation and calibration, (c) item storage and retrieval, (d) linking new items to the bank, and (e) maintenance of an item bank.

3.1   Item writing

The item writing process is a critical part of item bank development which requires both talent and skill. Unless there are a large enough number of well-trained and talented item writers, development of a calibrated item bank could not be efficient since the mortality rate of items in the validation processes may be high. Recruitment and training of item writers is clearly not an easy task, nor is it cheap. As an illustration, the following is Indonesia's scenario for the development of a national network of calibrated item banks for selected subjects taught in the lower and upper secondary schools. There are six subjects in the lower secondary school and eight subjects in the upper secondary school. For the continuity of item construction, it is planned that in each of the 27 provinces there should be at least three reasonably well-trained item writers. This means that 1,134 writers had to be recruited and trained; each of them attended a full week of intensive training on item writing using guidelines developed by the Examination Development Center. It took more than a year to implement such a training program even though courses were held twice a month with an average of 40 participants each. The cost was also enormous especially for travel since the locations are widely spread. Each item writer is assigned to construct ten items per month, hence, there are $10 \times 3 \times 27$ or 810 new items for each subject every month. In this case, each item writer is also assigned as an item reviewer for items from other provinces. Under this scenario, it is expected that 9,720 items could be collected every year. The cost of item writing is also high even with a low cost rate, relative to many industrialized countries. This project showed that more than half of the items being collected needed revision, and that some of them even had to be dropped. In conclusion, item writing for a relatively large-scale item bank is expensive and it is more an art than simply the mastery of item writing methodology and subject matter.

The practice of item writing for item banking may vary from one place to another. Mostly, the differences are regarding (a) who constructs the items, (b) how many times the items need to be reviewed and by whom, and (c) what aspects of an item need to be reviewed. Although algorithms for generating items using a computer are available (Millman, 1984), their applicability is still limited to a few content domains. In most cases, items are constructed either by teachers or subject matter specialists. It is also important to note, that sometimes there are disagreements among educators regarding what constitutes a good item. An item which is considered as good for one purpose might not be so for other purposes.

3.2 *Item validation and calibration*

It is mentioned above that item banking and Item Response Theory (IRT) are almost inseparable. Empirical validation of items in an item banking context is mostly verification of the extent to which item behavior follows a chosen IRT model. However, before empirical validation takes place, it is important to bear in mind that only good items in terms of both content validity and item construction criteria are subject to pilot testing; otherwise, a waste of resources and time would occur. The basic idea of the IRT is discussed below. Readers who are interested in the details of IRT, especially the technical aspects of it, could refer to Hambleton and Swaminathan (1985) or Hambleton *et al.* (1991) (see *Item Response Theory; Latent Trait Measurement Models*).

Apart from the mathematical complexities of the IRT procedure its basic idea is relatively easy to understand. It is a theory about how person variables together with item variables determine the response data when a person is responding to an item. Although there are many such variables, the theory assumes that only a few variables predominantly determine the response. In this case, it is believed that only one person variable which is the attribute to be measured by an item (e.g., proficiency in mathematics) and one or more item variables (such as its difficulty) are considered as most important. Since the individual values of both person and item variables included in the model are unknown and to be estimated from response data, and since the model is a probabilistic one, such variables are labeled as parameters of the model. Because there is only one person parameter, namely, ability and the performance on an item is believed to be dependent upon ability level, a curve showing the relationship between person ability and his or her performance on an item can be described. In this case, it is postulated that the curve is monotonically increasing along the ability continuum, which means the higher the ability the better the performance on the item. For a typical achievement test item with dichotomous response, the performance is represented by the correct answer. For a group

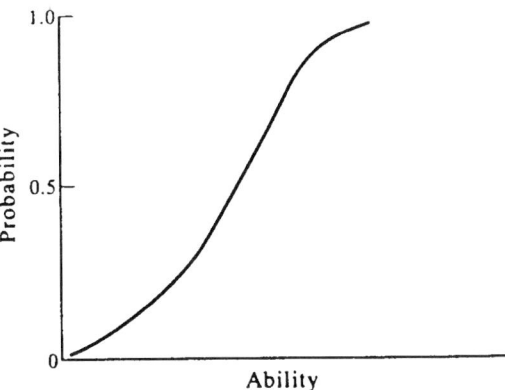

Figure 1   Theoretical item characteristic curve.

of persons of a given ability level, it is represented by the proportion or probability of correct answers for that group. Since the curve describes the characteristic of an item at different levels of ability, it is called an item characteristic curve (ICC). Figure 1 shows an ICC as postulated by the theory.

Empirical validation of items under IRT is basically a verification of the extent to which an item has an empirical ICC similar to that postulated by the IRT. To obtain this information, a set of items should be tried by a large number of respondents with heterogeneous ability, and a mathematical function describing the ICC should be chosen. In this way, estimation of the values of the parameters of the ICC for each item can be made, hence, the discrepancy between the empirical and theoretical curves can be measured (for each item). Figure 2 shows the empirical curve for two items and the theoretical curve. Item 1 is considered congruent to the expected curve.

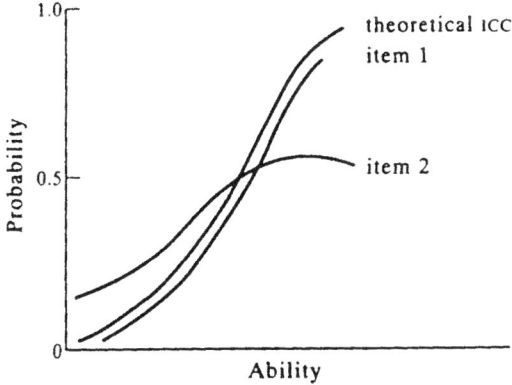

Figure 2   Empirical ICC of two items plotted against its theoretical ICC.

The most widely used mathematical form in describing ICC is the logistic function, which is a monotonically increasing function forming an S-shape curve. Another alternative is to use a normal ogive function. Models in IRT are usually labeled according to the number of item parameters involved. ICC for the one-parameter logistic model are given by the equation

$$P_i(\theta) = \frac{\exp(\theta - b_i)}{1 + \exp(\theta - b_i)} \quad i = 1, 2, \ldots, n \tag{1}$$

where $P_i(\theta)$ is the probability that respondents with ability θ answer item i correctly; b is the item difficulty parameter; and n is the number of items in the test. The equation for the two-parameter logistic model is

$$P_i(\theta) = \frac{\exp[1.7a_i(\theta - b_i))]}{1 + \exp[1.7a_i(\theta - b_i)]} \quad i = 1, 2, \ldots, n \tag{2}$$

where the additional item parameter is a, which is the item discrimination parameter. The constant 1.7 is needed to make the curve as close as possible to a normal ogive curve, which is the statistical basis of the two-parameter ICC. Another ICC model is the three-parameter logistic model whose equation is $P_i(\theta) = c_i + (1 - c_i)P_2(\theta)$ where c_i is the pseudochance-level parameter and $P_2(\theta)$ is the $P_i(\theta)$ in the two-parameter ICC.

Among the three models, the one-parameter logistic model, which is known as the Rasch model, is the most popular in item banking. It is in fact the most restricted one among the three models. However, there are some important features which are available only under the Rasch model. One of them is the possibility of estimating item parameters independently from the person's parameters, and vice versa. This provides not only better estimation of parameters even with a relatively small sample of examinees, but also provides invariance of item difficulty ordering regardless of the ability levels of examinees. This feature gives a clear and simple interpretation of the item difficulty parameter, and the concept of item calibration can be explained more readily. Under the two-parameter model, ICCs may cross each other so that one item could be easier than another for a particular ability group but the reverse might be observed in another group of examinees.

Calibration could be defined as the construction of a scale measuring the difficulty level of items on which the location of each individual item is to be determined. The scale construction involves two main activities, namely: (a) determining the scale's origin, and (b) defining/choosing a scaling unit. The concept of calibration may be shown in a simple way if one considers a set of items which fit the Rasch model. Since under the Rasch model items may vary only in their difficulty, such a set of items will have parallel ICCs as shown in Figure 3.

The difficulty level of an item can be defined as the point on the scale associated with the 50 percent chance for correctly answering that particular item. Items with ICCs on the

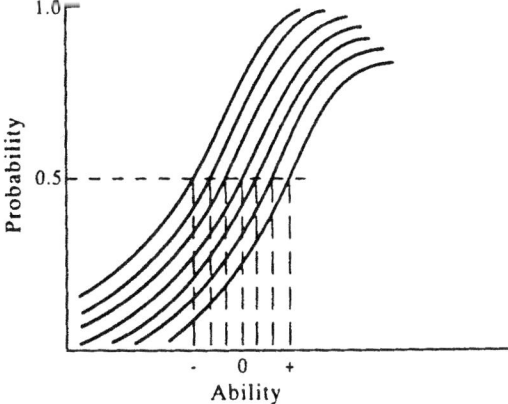

Figure 3   ICCs for a set of items which fit the Rasch model.

right side are more difficult than the ones on the left. Calibration in this case is carried out by first defining the origin of the scale (zero point), which is arbitrary. This can be done either by defining the position of any item as the zero point, or it can be defined statistically. The second step is to determine/measure distance between the location of each item from the zero point using scaling units as defined by the mathematical formulation of the ICC. Items to the left side of the zero point have negative values with a magnitude proportional to its relative distance from the zero point. Similarly, items on the right side of the zero point have positive values. Development of a calibrated item bank is basically keeping the existing scale already defined, while new items are added continuously into the bank. An important point to note is that only fitted items are subject to calibration. Mathematical procedures for calibrating items are explained in IRT text books such as Hambleton *et al*. (1991) or Wright and Stone (1979). A list of IRT computer software can be found in Hambleton *et al*. (1991).

3.3   *Linking new items*

As new items are continuously added to the bank, linking the scales obtained from new calibrations to the existing scale in the bank is an important part of item bank development. There are two important issues to be considered in this case: (a) design for ensuring a high quality and efficient link, and (b) estimation of the linking constant.

The most popular designs for linking new items to the existing bank are either using "common items" or "common persons". These designs are essentially the same because the scale for item difficulty is the same for measuring person ability. However, the common item approach is usually preferable because under this design a respondent would typically take fewer items than would be necessary under the common persons design. Figures 4a and 4b show two simple ways of linking the scales obtained from five sets of items administered to five different groups of examinees (darkened areas are

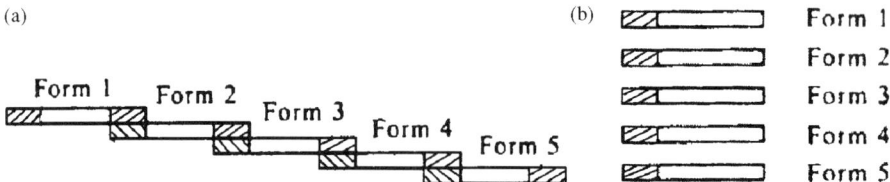

Figure 4 (a) Four sets of common items connecting five test forms. (b) One set of common items connecting five test forms.

common items). Interested readers may refer to Wright and Stone (1979) for many possible forms of linking design.

When the Rasch model is applied, the scales obtained from separate calibrations will differ only in its origin. Hence, links can be made by simply finding a constant representing distance between origins of two different scales. Here, the scales can be transformed from one to another by simple addition or subtraction using the constant. Under the Rasch model, this linking constant can easily be estimated by calculating the difference between the means of the difficulty estimates of the common items obtained from the two groups in which they were piloted. The number of common items is the crucial part of this linking design; the greater the number of common items the better, although fewer items could be calibrated. However, if the number of common items is too small, the link might not be reliable. There have been several studies in this field but with no agreement among the results. McKinley and Reckase (1980) recommended 15 items if a concurrent calibration is made, while Wingersky and Lord (1984) recommended as few as five common items. Umar (1987) showed that five common items are acceptable and ten common items are sufficient under the Rasch model. The same research also found that the simple linking estimates under the Rasch model are quite robust against violation of the Rasch model's assumptions.

Under the two-parameter model, the linking procedure is a little more complex. Here, the scales obtained from two separate calibrations will not only differ in origin but also in scaling unit. Linking equations for both item difficulty and item discrimination under this model, complete with good numerical examples, can be found in Hambleton *et al.* (1991).

4   Management and operations of an item bank

It is mentioned above that item banking is dependent upon information systems. The storage, cataloguing and retrieval of items clearly require computerization, particularly in larger banks. The computation involved in parameter estimation, designing the optimal test form for a particular purpose and test scoring, taking into account the known item parameters, would otherwise be impossible. In this case, the availability of good and easy to use software is very important for practical implementation of item banking.

There are three kinds of computer software needed in item banking practice. First, a database program suitable for storage, query, retrieval, formatting a test-page layout and printing the test paper. Ordinary database programs cannot handle achievement test items. Many items require graphical data to be inserted through either freehand drawings or optical scanning, while some other items contain mathematical or scientific symbols which are not available on the computer keyboard. In the selection of items to satisfy particular test specifications item data including the graphs and symbols should be arranged and displayed so that each item appears as on printed paper. Therefore, software specifically developed for item banking operation and management is required, however such software is still rare. Some institutions that have established and operated item banks, develop their own computer programs which are customized to their own needs so that others may have difficulty in using it without modifications.

The second type of computer software needed in item banking is statistical software for the estimation of IRT parameters and classical item analysis. Software of this type is readily available in the scientific software market but the potential users are limited since it requires technical knowledge in advanced statistics (particularly IRT). It is possible that there are institutions interested in the idea of item banking but which did not develop them due to lack of expertise in IRT.

The third type of software required by item banking is a tool for test scoring, a creative way of reporting test results, and for designing a test comprising the best possible combination of available items in the bank. Software for computerized adaptive testing can be considered to belong to this category. This type of software is usually user friendly but quite difficult to develop. Some institutions with a high level of technical expertise have developed prototype software of this type.

In addition to the computer, there are two subsystems which need to exist in item banking: a system of item production (including calibration and maintenance of items) and a system of utilization/services. For item production, it is necessary to have a continuous program of activities, carried out by full-time professionals, with an allocated budget, and using a tight schedule. It should be well-organized as opposed to an *ad hoc* and incidental activity.

A system of utilization or services should be developed if item banking is to be beneficial to an educational system. An example of a well-designed testing service system utilizing an advanced item bank can be seen in van Theil and Zwarts (1986). Establishment and maintenance of a sophisticated item bank is costly. In order to make the system efficient, optimal utilization should be achieved.

5   Problems in item banking

Since item banking is almost inseparable from IRT and computers, most problems that arise in item banking are associated with them, either directly or indirectly. The rejection of item banking is often based on reservations and the rejection of the application of IRT or the computer. Critics of the use of IRT could be best answered by efforts to develop further such a theory so that it could cope with situations that it cannot currently handle. The development of partial credit scoring under the Rasch model (Masters, 1982; Masters and Evans, 1986), for example, has made item banking more acceptable and beneficial because some important types of item are no longer excluded from its operations. Latest developments in computer software and hardware could also make item banking practices easier.

At the time of writing in the late-1990s, the following problems arise in the practical implementation of item banking.

(a) Item banking requires an expensive investment especially in the beginning.
(b) Item banking requires highly specialized professionals. This is also expensive.
(c) There is a lack of powerful but easy to use software that could make advance applications by nonspecialists possible.
(d) The construction of items satisfying the IRT models is difficult. The more restricted the IRT model in use, the greater the likelihood of rejecting items, including the ones that might be good for particular purposes.
(e) The requirements of IRT are sometimes difficult to satisfy, especially for achievement test items in the social sciences.

6   Conclusion

Despite the various ways of defining an item bank, teachers and test developers could greatly benefit from establishing and utilizing a large collection of achievement test items. It can also be shown that a calibrated item bank provides flexibility, efficiency, quality of items, testing security, measurement consistency and facilitates criterion-referenced evaluation. The recruitment and training of item writers and item reviewers is important but expensive, and the establishment of an item banking system should include systems of item production, management and operation (computer systems), and a system of utilization services. Finally, future research on item banking should be directed toward overcoming the existing problems so that its practice could be more beneficial to educational systems and more acceptable to teachers and educators (see also: *Adaptive Testing*).

References

Baker, F. B. 1986. Item banking in computer-based instructional systems. *Appl. Psychol. Meas.* **10**(4), 405–14.
Brzezinski, E. J. and Hiscox, M. D. (eds.) 1984. Microcomputers and testing. *Educational Measurement: Issues and Practices* **3**, 4–34.
Choppin, B. H. 1976. Item banking development. In: De Gruijter, D. N. M. and van der Kamp, L. J. T. (eds.) 1976. *Advances in Psychometrics and Educational Measurement*. Wiley, London.
Choppin, B. H. 1981a. Educational measurement and the item bank model. In: Lacey, C. and Lawton, D. (eds.) 1981. *Issues in Evaluation and Accountability*. Methuen, London.

Choppin, B. H. 1981b. Principles of Item Banking (Unpublished training material).

Goldstein, H. 1981. Limitations of the Rasch model scale for educational assessment. In: Lacey, C. and Lawton, D. (eds.) 1981. *Issues in Evaluation and Accountability.* Methuen, London.

Hambleton, R. K. and Swaminathan, H. 1985. *Item Response Theory, Principles and Applications.* Kluwer-Nijhoff, Boston, Massachusetts.

Hambleton, R. K. 1986. The changing conception of measurement: A commentary. *Appl. Psychol. Meas.* **10**, 415–21.

Hambleton, R. K., Swaminathan, H. and Rogers, H. J. 1991. *Fundamentals of Item Response Theory.* Sage, Newbury Park, California.

Masters, G. N. 1982. A Rasch model for partial credit scoring. *Psychometri.* **47**, 149–74.

Masters, G. N. and Evans, J. 1986. Banking non-dichotomously scored items. *Appl. Psychol. Meas.* **10**, 355–67.

McKinley, R. L. and Reckase, M. D. 1980. *A Successful Application of Latent Trait Theory to Tailored Achievement Testing.* Research Report 80-1. University of Missouri, Columbia, Missouri.

Millman, J. 1984. Individualizing test construction and administration by computer. In: Berk, R. A. (ed.) 1984. *A Guide to Criterion-referenced Test Construction.* Johns Hopkins University Press, Baltimore, Maryland.

Popham, W. J. 1978. *Criterion Reference Measurement.* Prentice-Hall, Englewood Cliffs, New Jersey.

Roid, G. H. and Haladyna, T. J. 1982. *A Technology for Test-item Writing.* Academic Press, New York.

Umar, J. 1987. Robustness of the simple linking procedure in item banking using the Rasch model. (Doctoral dissertation, University of California, Los Angeles).

Umar, J. 1990. Development of an examination system based on calibrated item bank networks. Unpublished Project Report, SIDEC, Stanford University, Stanford, California.

Van der Linden, W. J. 1986. Forewords to the special issues in item banking. *Appl. Psychol. Meas.* **10**(4).

Van Theil, C. C. and Zwarts, M. A. 1986. Development of a testing service system. *Appl. Psychol. Meas.* **10**, 371–403.

Wingersky, M. S. and Lord, F. M. 1984. An investigation of methods for reducing sampling error in certain IRT procedures. *Appl. Psychol. Meas.* **8**, 347–64.

Wright, B. D. and Stone, M. 1979. *Best Test Design.* MESA Press, Chicago, Illinois.

Further reading

1984 Special issues in item banking. *J. Educ. Meas.* **21**(4).

1986 Special issues in item banking. *Appl. Psychol. Meas.* **10**(4).

Roid, G. H. 1989. Item writing and item banking by micro-computer: An update. *Educational Measurement: Issues and Practices* **8**(3), 17–20.

18 Item Bias

J. D. Scheuneman and C. A. Bleistein

Item bias procedures are used to determine whether the individual items on an examination function in the same way for two groups of examinees, usually defined by racial and ethnic background, sex, age and experience, or condition of handicap. These procedures would be applied when the examinee groups of interest appear to differ in their mean level of the ability, knowledge, or skill being measured, making a direct comparison of their performance on the items inappropriate. In most instances these methods are applied when no criteria of the abilities, knowledge, or skills being measured are available outside of the examination in question.

Although the test performance of various population subgroups has long been of interest, the issue of the fairness of the use of tests with these subgroups came into prominence in the measurement literature in the late 1960s and early 1970s. A number of procedures were proposed and evaluated, but all required some criterion of performance in the college or job setting, the two situations in which these methods were most frequently applied.

At the same time, test publishers became interested in instituting procedures to identify problematic items, which could then be revised or eliminated before the test forms were finalized. Criterion measures could not be obtained until tests were in use, however, making the existing procedures inapplicable. Consequently, a number of procedures were devised which were useful where external criteria were unavailable.

Although not all of the procedures that have been suggested have stood the test of time, a number of methods are now generally accepted as useful in isolating items that appear to function differently for two groups. Such methods are now included as part of the test development process of a number of tests (Berk, 1982; Diamond and Elmore, 1986). Notice, however, that, just as the test bias procedures are inapplicable in this setting, the item bias procedures are not appropriate for evaluating bias or fairness in test use.

For the purpose of this entry, the various methods for detecting item bias are divided into two categories. In the first of these categories are methods based on observed item responses and test scores that use classical measurement methods. The second category contains those approaches based on "true" abilities, as in item response theory models and methods. Throughout the entry, the term "item bias" has been replaced by "differential item functioning" (DIF). This term is coming to be preferred by many researchers because of its focus on what is actually observed rather than suggesting inferences about the nature of the effect. (See Scheuneman, 1982, for a discussion of this point.)

1 Classical approaches

1.1 Transformed item difficulty methods

Essentially these methods of detecting DIF involve calculating percent correct item difficulties (*p*-values) for two groups of examinees and transforming the *p*-values to another metric in order to make the relationship between the respective difficulty values linear. The most widely used of these procedures is the delta plot method (Angoff and Ford, 1973). In this procedure, *p*-values are converted to deltas; that is normal deviates with a mean of 13 and a standard deviation of 4. Typically, a plot on a bivariate chart of the deltas for two groups of equivalent ability will form an ellipse along the 45-degree line through the origin which would represent equal difficulty of the items. When the groups being compared differ in ability, the points will be displaced from this line and the ellipse may be rotated somewhat from 45 degrees. See Figure 1 for an example of a delta plot.

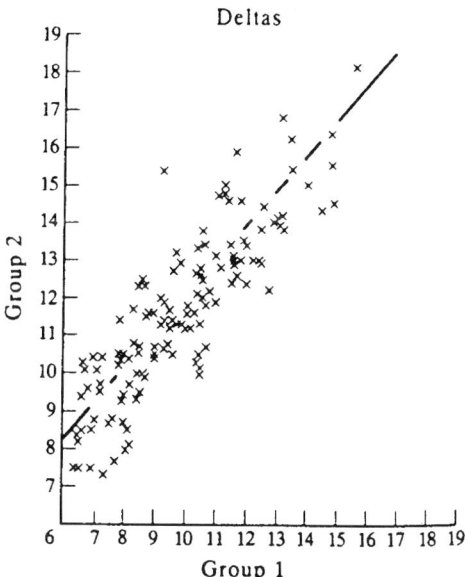

Figure 1   Example of transformed item difficulties (deltas) plotted for two groups.

The major axis of the ellipse is defined in terms of the points representing the item difficulties for the two groups. Using X and Y to represent the respective item difficulties for the two groups being compared, the major axis is given by:

$$Y = aX + b \qquad (1)$$

where:

$$a = \frac{(S_y^2 - S_x^2) \pm \sqrt{(S_y^2 - S_x^2)^2 + 4r_{xy}^2 S_x^2 S_y^2}}{2r_{xy} S_x S_y}, \qquad (2)$$

and

$$b = M_y - aM_x \qquad (3)$$

The distance of each item from the major axis is measured by:

$$d_i = (ax_i - y_i + b)/a^2 + 1 \qquad (4)$$

Other indices used with the delta plot method include the standard deviation of d_i and the correlation between the delta values of the two groups, the magnitude of which reflects the degree to which items rank in the same order of difficulty in both subgroups. The method may also be generalized to the investigation of several groups simultaneously (Angoff, 1975). Modifications have been proposed by Sinnott (1980) and Rudner *et al.* (1980a).

Advantages of the delta plot method are that it is simple, inexpensive, easily explained and does not require large numbers of examinees, although a reasonably large set of items is required to ensure that the major axis is well-defined. A principal disadvantage is that when two groups differ in their mean ability, an item that is unusually discriminating will result in larger item difficulty differences, while an item that has particularly low discrimination will show smaller differences than the other items on the test even when the items are not functioning differentially. Angoff (1982) suggested using groups matched on ability to circumvent this problem. The method is sample dependent, as is the case for all the classical procedures, and the delta values for different-sized samples are not equally reliable. This latter difficulty can be overcome by using an arcsin transformation rather than the inverse normal transformation used by the delta plots (Cardall and Coffman, 1964; Plake and Hoover, 1979–80).

1.2   Item discrimination procedures

Pairs of item discrimination indices (point-biserials) for different ethnic or gender groups have been plotted in the manner described for transformed item difficulties and the plots examined for outliers in an attempt to identify differentially functioning items. Green and Draper (1972) used point-biserial correlations to separate the items in a test into "best half" and "worst half" and defined differentially functioning items as those chosen as best for one group and worst for another. These methods have the same advantages as the transformed item difficulty methods. Unfortunately, studies of these procedures have suggested that these methods lack validity for identifying items that function differentially (Ironson and Subkoviak, 1979; Rudner, 1977 and Shepard *et al.*, 1981).

1.3 Contingency table approaches

The contingency table methods were first suggested by Scheuneman (1979) based on a definition stating that for an item that does not function differentially, persons of equal ability have equal probability of a correct response regardless of their group membership. This definition was operationalized in a contingency table made up of group membership and ability, which was defined by score ranges on the test or subtest on which the item appeared. A number of variations of this technique have since been suggested and are discussed below. All of these methods assume that the test score, which is used as the ability measure, is valid and reliable and that the test is essentially unidimensional. Without unidimensionality, it cannot be assumed that two people with the same score have equivalent ability.

1.3.1 Chi-square methods

Scheuneman's (1979) contingency table method and variations which use a chi-square value as an index are collectively referred to as chi-square methods. After collapsing the score scale into J intervals (typically 3 to 5), an index value is computed. Scheuneman's original index, C2, which is based only on correct responses, is calculated as follows:

$$C2 = \sum_{j=1}^{j} \frac{(E_{j1} - O_{j1})^2}{E_{j1}} + \sum_{j=1}^{j} \frac{(E_{j2} - O_{j2})^2}{E_{j2}} \tag{5}$$

where:

O_{jk} = the observed frequency of correct responses for score interval j and group k:
$E_{jk} = P_{j.} \cdot N_{jk}$ = the expected frequency of correct responses for score interval j and group k;
$P_{j.}$ = proportion correct across all groups within score interval j;
N_{jk} = the number of examinees in score interval j and group k.

The procedure may be readily extended to evaluate item functioning for several groups simultaneously. Not only does this provide computational efficiency, but the results provide information on the relative standings of the groups that is not available from pairwise comparisons. For example, if four groups are being compared and the item with a high index value is functioning similarly for three of them, it is clear for which group the item is functioning differentially and whether this group is favored or disfavored by the item.

Scheuneman's method has been criticized because the associated index is not distributed as the chi-square, although it has a chi-squarelike appearance. Hence it does not have an associated sampling distribution and significance tests cannot properly be employed. The modification that has most commonly been suggested includes the incorrect responses in the contingency table, a variation often attributed to Camilli (1979). It is possible to derive a formula including this information which permits the chi-square index to be obtained without actually having to compute the expected values for the incorrect responses (Scheuneman, 1981). This formula is as follows:

$$\chi^2_{\text{full}} = \sum_{j=1}^{j} \frac{(E_{j1} - O_{j1})^2}{E_j(1 - P_{j0})} + \sum_{j=1}^{j} \frac{(E_{j2} - O_{j2})^2}{E_{j2}(1 - P_{j0})} \tag{6}$$

The chi-square methods have several important advantages: notably their intuitive appeal due to the use of groups matched on ability, simplicity and appropriateness for small sample sizes. They are also inexpensive to obtain. The C2 and the full chi-square index tend to produce very similar results (Scheuneman, 1986), but can be contrasted on a number of points. When sample sizes are large, the full chi-square has tended to perform somewhat better in research studies, but minimum samples are probably about 300 in contrast to about 100 for the C2 index. The full chi-square does better in evaluating difficult items; the C2 does better for easy items. Both methods have been criticized because groups may not be equivalent in ability within the broad score intervals used.

1.3.2   Log–linear methods

General log–linear models provide a means of analyzing qualitative data through their relationship to the elements of contingency tables. The application of these models to the evaluation of DIF involves: (a) the construction of a three-way contingency table (ability level by group by item response) for each item similar to that used with the chi-square methods; (b) specification and fitting of the models of interest; (c) calculation of a residual goodness-of-fit measure, such as the chi-square likelihood ratio (G^2); and (d) testing for significant differences between models; that is the difference between the G^2 obtained for the two models.

Three hierarchical models are typically used in the study of DIF. The first tests for a main effect of ability level; in the second, a term is added for a main effect of group; and the last requires a term for the interaction of ability with group. If Model 1 adequately fits the data, no DIF exists. If Model 2 provides a significantly better fit, the item is exhibiting DIF that is uniform across ability levels; and if Model 3 is needed to explain the data, the DIF is nonuniform in nature (see Mellenbergh, 1982).

Examples of research applying this methodology to the detection of DIF include the work of Alderman and Holland (1981), Mellenberg (1982) and van der Flier et al. (1984). An advantage of log–linear models over the other contingency table methods is that they provide added information by distinguishing between uniform effects, resulting where the differential functioning is with regard to difficulty only, and nonuniform effects, resulting where differences in item discrimination also exist. They may also be used with relatively small sample sizes. The major disadvantage is that the available software uses an iterative algorithm to process the data so the method is more expensive than the other classical procedures and requires more expertise to use.

1.3.3   Mantel–Haenszel procedure

This procedure, developed by Mantel and Haenszel (1959), is closely related to log–linear procedures for detecting first-order interactions, and has been used extensively in biomedical research. Its application to the study of DIF was introduced by Holland (Holland and Thayer, 1988). As with the other contingency table methods, performance is compared for group members of comparable ability, with total test score as the matching criterion. The Mantel–Haenszel (MH) estimate (α_{MH}) is a weighted average of the odds ratio at each of j scores, and may be interpreted as the average factor by which the likelihood that a member of one group answers the item correctly exceeds the corresponding likelihood for a member of the other group. The equation for the MH estimate is:

$$\alpha_{MH} = \left(\sum_j A_j D_j / T_j \right) \Big/ \left(\sum_j B_j C_j / T_j \right) \tag{7}$$

where:

  A_j = the number of group 1 members responding correctly;
  B_j = the number of group 1 members responding incorrectly;
  C_j = the number of group 2 members responding correctly;
  D_j = the number of group 2 members responding incorrectly;
  $T_j = A_j + B_j + C_j + D_j$ = the total number of examinees with score j who responded to the
    item.

An α_{MH} value of 1.00 indicates that a correct response is equally likely for both groups. If α_{MH} is greater than 1.00, group 1 members are more likely to respond correctly, and if less than 1.00, group 2 members have the advantage. To test the hypothesis of independence of response and group membership, a one-degree of freedom chi-square test is associated with the MH estimator and is calculated as:

$$\chi^2_{MH} = \frac{\left(\left| \sum_j A_j - E(A_j) \right| \frac{1}{2} \right)^2}{\sum_j S(A_j)^2} \tag{8}$$

where:

$$E(A_j) = [(A_j + B_j)(A_j + C_j)]/T_j. \tag{9}$$

and:

$$S(A_j)^2 = \frac{(A_j + B_j)(C_j + D_j)(A_j + C_j)(B_j + D_j)}{T_j^2(T_j - 1)} \tag{10}$$

A major difference between the MH procedure and the contingency table methods discussed so far is that it matches on total test score by unit intervals, thus avoiding the problems associated with collapsing the score scale into categories. Sample sizes required are therefore greater than those required for the previously discussed methods (about 500 per group), but smaller than those for the three-parameter item response theory models to be discussed below. In addition, the MH procedure does not require computer programs using expensive iterations as do log–linear and item response theory models. If nonuniform DIF is of concern, MH may not be the best choice of method since it may fail to detect instances of this type of DIF.

1.4 Standardization procedure

The standardization procedure, developed by Dorans and Kulick (1986), is based on an empirical item response function where the probability of a correct response to an item is estimated by the observed proportion correct at each ability level (typically measured in unit intervals of total test score). Estimates of the conditional probability of success at each score level are developed on the "reference group", the group that forms the performance reference for the "focal group" (the group of interest, which is typically the lower scoring group). The reference group is usually the larger sample and provides the most stable estimates across the score range.

The DIF index provided by the standardization method uses a weighting function supplied by the standardization group. In practice, the weight chosen has been the number of focal group examinees at a given score level (K_j) because this weights the differences between the probabilities of a correct response for the reference and focal groups most heavily at score levels most often achieved by the focal group members. However, other options are available for the weighting function. The item discrepancy index is the standardized p-difference (D_{STD}), defined as:

$$D_{STD} = \sum_j K_j(P_{if} - P_{jr}) \Big/ \sum_j K_j \tag{11}$$

where

$(K_j/\Sigma\, K_j) =$ the weighting factor;

P_{if} = the probability that a focal group member at a given score level will answer the item correctly;

P_{jr} = the corresponding probability for the reference group.

An advantage of the standardization procedure is that visual aids have been developed to assist in interpretation. Graphical displays of the conditional probabilities of successful performance and the difference between these probabilities for focal and base groups for each item are shown in Figures 2a and 2b. Typically, these plots are produced only for items flagged for high values of D_{STD}.

Computationally the standardization and Mantel–Haenszel approaches are very similar, with the primary difference being in the weights used, although the standardization estimates tend to be more stable. The principle disadvantage of both methods is that large samples are required for stable estimation. (See Dorans, 1989, for a comparison of these two procedures.)

1.5   Logistic regression method

Conceptually related to the contingency table methods, particularly the log–linear method, the logistic regression method is designed to detect both uniform and nonuniform DIF (Swaminathan and Rogers, 1990). Rather than treating the ability dimension as a set of unordered categories, however, this procedure retains ability as a continuum. In the model, the probability of a correct response is a function of z, where:

$$Z_i = \tau_0 + \tau_1 \theta + \tau_2 g + \tau_3 (\theta g) \tag{12}$$

and $\tau_0 \tau_1$ = intercept and slope of the logistic function;

θ = examinee ability;

$g = 1$ for examinees belonging to group 1 and 0 for examinees belonging to group 2;

θg = the product of g and θ.

The parameter τ_2 corresponds to the group difference in performance and τ_3 to the interaction between group and ability. An item shows uniform DIF if $\tau_2 \neq 0$ and $\tau_3 = 0$, and nonuniform DIF if $\tau_3 \neq 0$, whether or not $\tau_2 = 0$. The values of τ_1, τ_2, τ_3 and τ_4 are estimated using maximum likelihood methods. Significance tests are available.

This method has not yet been widely evaluated but appears to compare well with the Mantel–Haenszel procedure for simulated instances of uniform bias, and will detect instances of nonuniform bias that the MH procedure misses. It is more difficult to compute, however, and, because it uses iterative procedures, is more expensive than MH and many of the other classical procedures.

1.6   Distractor analysis

In addition to flagging items for DIF, distractor analysis may be used to help provide an explanation of the result. Scheuneman (1982) described three methods for analysis of distractors: (a) her comparison of mean-scaled scores of individuals responding to each option, (b) Veale and Foreman's (1983) chi-square based on the incorrect options only, and (c) Frary and Giles's (1980) Rasch model method. Green *et al.* (1989) propose the use of log–linear models to examine the possibility of a subgroup-by-option interaction when ability is held constant. They advocate analyzing DIF through incorrect responses and describe their method as the study of differential distractor functioning. Any of the

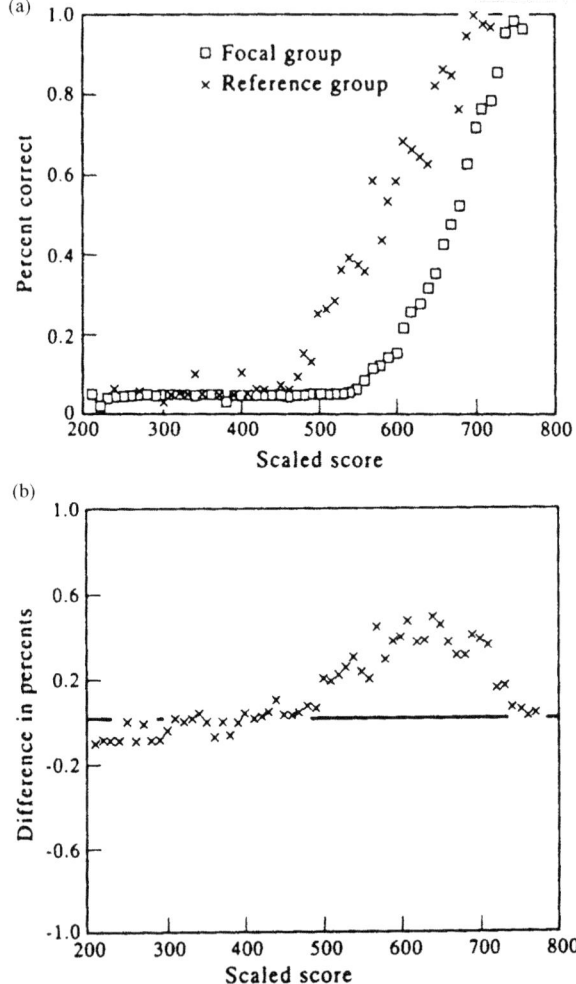

Figure 2   (a) Conditional probabilities of successful performance for reference and focal groups. (b) Difference in conditional probabilities of successful performance between reference and focal groups.

contingency tables methods can also be used to analyze distractors, where sample sizes are large enough, by repeating the procedures on one or more of the distractors rather than on the keyed response.

2   Item response theory approaches

The DIF procedures using item response theory (also known as latent trait theory) are often considered preferable to the classical approaches discussed above because of their theoretical underpinnings. The assumption of item response theory (IRT), that makes it useful for investigating DIF, is that the estimated parameters of the item response function (IRF) are invariant for different samples drawn from the same population (see *Item Response Theory*). Hence, if parameters are estimated separately for two groups, the resulting IRFs of an item which is functioning equivalently for those groups should be the same (apart from a linear transformation of scale). This means that the probability of a correct response for persons at a given ability level is the same for both groups, a statement equivalent to the definition given for the contingency table methods except that the true ability scale is used rather than observed test scores. Where there is DIF, however, the IRFs will be displaced from each other somewhat. Figure 3 shows the IRFs for two such items.

2.1   Three-parameter methods

Although all of the IRT methods agree that an item for which the IRFs are different for two groups is functioning differentially for those groups, a number of procedures for distinguishing such differences from error of parameter estimation have been suggested. As yet, no clear preference for any one of these procedures has emerged.

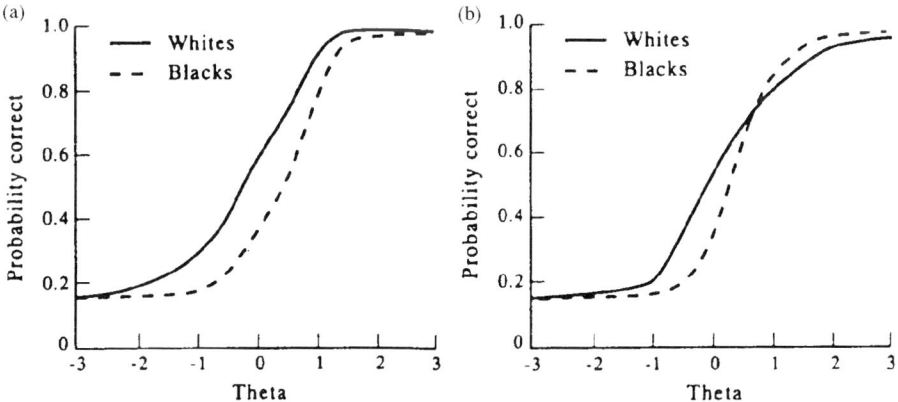

Figure 3   IRFs obtained for two items that demonstrate DIF for Black and White examinees.

In the method proposed by Rudner (1977), the items are calibrated separately for each of the two groups being compared and equated for scale. The area between the two obtained curves is then approximated by summing the difference between the probability of a correct response at small increments of ability. The formula for the area is expressed by:

$$\varphi_k = \sum_{-4.0}^{4.0} |P_1(u_k = 1 | \theta_j) - P_2(u_k = 1 | \theta_j)| \Delta\theta \tag{13}$$

where $P(u_k = 1) =$ the logistic function with the parameter values that were obtained when calibrating the item for group k, and $\Delta\theta = 0.005$.

Once the item calibrations have been carried out, this statistic can be calculated readily with a simple computer program. Perhaps for this reason it is the method used in most of the research comparing the three-parameter IRT methods to other DIF procedures. Raju (1990) has developed significance tests for both signed and unsigned area measures, thus enhancing the usefulness of this method.

For the procedure suggested by Lord (1980), items are first calibrated with all examinees together in order to get better estimates of the c parameters. These are then held constant and the a and b parameters are re-estimated for the two groups separately and equated for scale. Lord then suggests a chi-square test which permits simultaneous comparison of the a and the b parameters in the two separate calibrations. The significance test is asymptotic; it assumes that the abilities are known, rather than estimated; and can be applied only to maximum likelihood estimates of a and b. The primary disadvantage of this method is that it requires a complex computer program to obtain the chi-square values, although this is not particularly expensive to run once the parameter estimates have been obtained.

Linn and Harnish (1981) suggested a procedure that compares the actual performance of focal group members with that predicted by the model. The items are first calibrated on the total group and estimates of ability are obtained for the focal group. Fit is then assessed with a standardized difference score using the following formula:

$$Z_i = \frac{1}{N_f} \sum_{cf} \frac{U_{ij} - P_{ij}}{P_{ij}(1 - P_{ij})} \tag{14}$$

where:

$U_{ij} = 1$ if item i is answered correctly by person j, and 0 otherwise;
$N_f =$ the number in group f;
$P_{ij} =$ estimated probability that person j would answer item i correctly, based on the fitted model from combined groups.

This method permits the use of IRT methods in instances where the total sample is large enough for IRT-item calibration, but the focal group of interest is not. This is a common situation for practical applications of DIF methods. As with the area method, the statistics are easy to compute once the calibrations have been done. If the model does not fit the data for that item, either for the focal or reference group, however, DIF may be identified where it does not in fact exist.

In those instances where the IRT model can be shown to fit the data, these procedures offer a strong theoretical basis and the advantages of true ability estimates rather than observed scores. Research suggests that these methods are superior to the classical approaches for the identification of DIF (Ironson and Subkoviak, 1979; Rudner *et al.*, 1980b; Shepard *et al.*, 1981). In general, however, the three-parameter methods will be more expensive than either the classical methods or the Rasch model methods. LOGIST, one of the more frequently used computer programs for parameter estimation, is an iterative program and is consequently very expensive to run. Large samples are required (at least 1,000 people per group and 40 items), and the *c* parameters often fail to converge. Robustness of the estimation techniques to violation of the IRT model assumptions is unknown. (See Hills, 1989, for a summary of the advantages and disadvantages of the different IRT indices as well as several of the classical indices.)

2.2 The Rasch model

In the Rasch model, the discrimination of the items is assumed to be constant and the lower asymptote of the IRF is assumed to be zero (see *Rasch Measurement Theory*). Hence, if the IRFs for two groups differ, the difference must be in the location parameter, *d*, representing item difficulty. The two most common procedures used for evaluating DIF with the Rasch model examine either the differences in difficulty between groups (the difficulty shift), or the fit of each item to the model in each group (Draba, 1977; Durovic, 1975; Wright *et al.*, 1976).

To analyze a difficulty shift, the *d* parameters are estimated for each group separately and placed on the same scale. A *t* statistic may then be used to evaluate the difference in *d* parameters for the two groups for each item, where:

$$t_i = (d_{i1} - d_{i2})/(SE_{i1}^2 + SE_{i2}^2) \tag{15}$$

and *SE* represents the standard error of *d*. If *t* is large, the item is functioning differentially in the two groups.

Analysis of the fit of each item to the model assumes that a non-DIF item will have a similar fit in each group. Again, statistics are generated for each group separately and a chi-square with one degree of freedom is computed for each item in each group as:

$$\chi^2 = Z_{ij}^2 = \frac{[x_{ij} - E(x_{ij})]^2}{S^2(x_{ij})} \tag{16}$$

where:

x_{ij} = an individual's score ($i = 1$ or 0) on item j:
$E(x_{ij})$ = the expected response predicted by the model;
$S^2(x_{ij})$ = the variance of the expected value.

The overall fit of an item is given by the mean square:

$$MS_i = \sum_{j}^{n} Z_{ij}^2 [L/(n-1)(L-1)] \qquad (17)$$

where: L = test length and n = the number of individuals in the sample.

MS has an expected value of 1 and a standard error of $2L/(n-1)(L-1)$. Fit of each item to the model for different groups has also been assessed by the difference in mean square residuals (Wright *et al.*, 1976).

In their evaluation of DIF, Rudner *et al.* (1980b) found the difficulty shift analysis to be better than the fit statistic. Shepard *et al.* (1981) found a near perfect correlation of results between the difficulty shift statistic and a delta plot distance measure. Both studies found that the Rasch model methods worked less well than the chi square procedures.

The major advantages of the Rasch model are the theoretical parameter invariance and the availability of the computer programs to process the data inexpensively. There is disagreement regarding the minimum sample required. Some have argued that as few as 100 are adequate to obtain good estimates of the d parameter (Wright, 1977). The crucial problem for these techniques is whether the Rasch model provides an adequate fit to the data.

3   Prospects and challenges for research

Work will continue into the future on statistical or procedural refinements of the methods described above. New statistical methods may also emerge in the coming years. Two major challenges will need to be met in future research that go beyond the methodologies used in the early 1990s. First is the challenge set by modes of assessment other than multiple choice where items may not be dichotomously scored. Further, some of these assessment modes may not have an internal criterion measure of ability, such as a total score used with multiple-choice tests, that is adequate for the use of the DIF procedures as they are conceptualized. Second is the challenge raised by the frequent inability of researchers to move beyond the statistical results to an understanding of why these differences occur and what corrective actions – if any – might be taken.

References

Alderman, D. L. and Holland, P. W. 1981. *Item Performance Across Native Language Groups on the Test of English as a Foreign Language*. Educational Testing Service, Princeton, New Jersey.

Angoff, W. H. 1975. The investigation of test bias in the absence of an outside criterion. Paper presented at the National Institue of Education Conference on Test Bias, Annapolis, Maryland.

Angoff, W. H. 1982. The use of difficulty and discrimination indices in the identification of biased test items. In: Berk, R. A. (ed.) 1982.

Angoff, W. H. and Ford, S. F. 1973. Item–race interaction on a test of scholastic aptitude. *J. Educ. Meas.* **10**, 95–106.

Berk, R. A. (ed.) 1982. *Handbook of Methods for Detecting Test Bias.* Johns Hopkins University Press, Baltimore, Maryland.

Camilli, G. 1979. A critique of the chi square method for assessing item bias. Unpublished paper, Laboratory of Educational Research, University of Colorado, Boulder, Colorado.

Cardall, C. and Coffman, W. E. 1964. *A Method for Comparing the Performance of Different Groups on the Items in a Test.* Educational Testing Service, Princeton, New Jersey.

Diamond, E. E. and Elmore, P. B. 1986. Bias in achievement testing: Follow-up report of the AMECD Commission on Bias in Measurement. *Meas. Eval. Couns. Dev.* **19**, 102–12.

Dorans, N. J. 1989. Two new approaches to assessing differntial item functioning: Standardization and the Mantel–Haenszel method. *Applied Measurement in Education* **2**, 217–33.

Dorans, N. J. and Kulick, E. 1986. Demonstrating the utility of the standardization approach to assessing unexpected differential item performance on the Scholastic Aptitude Test. *J. Educ. Meas.* **23**, 355–68.

Draba, R. E. 1977. *The Identification and Interpretation of Item Bias.* Statistical Laboratory, Department of Education, University of Chicago, Chicago, Illinois.

Durovic, J. J. 1975. Test bias: An objective definition for test items. ERIC Document Reproduction Service No. ED 128 381, Washington, DC.

Frary, R. B. and Giles, M. B. 1980. Multiple choice test bias as reflected by examinee selection of inappropriate answers. Report submitted to the National Institute of Education under grant No. NIE-G-79-0140.

Green, B. F., Crone, C. R. and Folk, V. G. 1989. A method for studying differential distractor functioning. *J. Educ. Meas.* **26**, 147–60.

Green, D. R. and Draper, J. F. 1972. Exploratory studies of bias in achievement tests. ERIC Document Reproduction Service No. ED 070 794, Washington, DC.

Hills, J. R. 1989. Screening for potentially biased items in testing programs. *Educational Measurement: Issues and Practice* **8**(4), 5–11.

Holland, P. W. and Thayer, D. T. 1988. Differential item performance and the Mantel–Haenszel procedure. In: Wainer, H. and Braun, H. I. (eds.) 1988. *Test Validity.* Erlbaum, Hillsdale, New Jersey.

Ironson, G. H. and Subkoviak, M. J. 1979. A comparison of several methods of assessing item bias. *J. Educ. Meas.* **16**, 209–25.

Linn, R. L. and Harnish, D. 1981. Interactions between item content and group membership in achievement test items. *J. Educ. Meas.* **18**, 109–18.

Lord, F. M. 1980. *Application of Item Response Theory to Practical Testing Problems.* Erlbaum, Hillsdale, New Jersey.

Mantel, N. and Haenszel, W. 1959. Statistical aspects of the analysis of data from retrospective studies of disease. *Journal of the National Cancer Institute* **22**, 719–48.

Mellenbergh, G. 1982. Contingency table models for assessing item bias. *J. Ed. Stat.* **7**, 105–18.

Plake, B. S. and Hoover, H. D. 1979–80. An analytical method of identifying biased test items. *J. Exp. Educ.* **48**, 153–4.

Raju, N. 1990. Determining the significance of estimated signed and unsigned areas between two item characteristic curves. *Appl. Psychol. Meas.* **14**, 197–207.

Rudner, L. M. 1977. An evaluation of select approaches for biased item identification. (Doctoral dissertation, Catholic University of America).

Rudner, L. M., Getson, P. R. and Knight, D. L. 1980a. Biased item detection techniques. *J. Ed. Stat.* **5**, 213–33.

Rudner, L. M., Getson, P. R. and Knight, D. L. 1980b. A Monte Carlo comparison of seven biased item detection techniques. *J. Educ. Meas.* **17**, 1–10.

Scheuneman, J. D. 1979. A method of assessing bias in test items. *J. Educ. Meas.* **16**, 143–52.

Scheuneman, J. D. 1981. A response to Baker's criticism. *J. Educ. Meas.* **18**, 63–6.

Scheuneman, J. D. 1982. *A posteriori* analyses of biased items. In: Berk, R. A. (ed.) 1982.

Scheuneman, J. D. 1986. Differential item performance: Use of computer simulation to evaluate indices. In Angoff, W. H. (ed.) 1986. *Differential Item Performance: Methodological and Measurement Issues*. American Educational Research Association, San Francisco, California.

Shepard, L. A., Camilli, G. and Averill, M. 1981. Comparison of procedures for detecting test-item bias using both internal and external ability criteria. *J. Educ. Stat.* **6**, 317–75.

Sinnott, L. T. 1980. *Differences in Item Performance Across Groups*. Educational Testing Service, Princeton, New Jersey.

Swaminathan, H. and Rogers, H. J. 1990. Detecting differential item functioning using logistic regression procedures. *J. Educ. Meas.* **27**, 361–70.

Van der Flier, H., Mellenbergh, G., Ader, H. J. and Wijn, M. 1984. An iterative item bias detection method. *J. Educ. Meas.* **21**, 131–45.

Veale, J. R. and Foreman, D. I. 1983. Assessing cultural bias using foil response data: Cultural variation. *J. Educ. Meas.* **20**, 249–58.

Wright, B. D. 1977. Solving measurement problems with the Rasch model. *J. Educ. Meas.* **14**, 97–115.

Wright, B. D., Mead, R. J. and Draba, R. 1976. *Detecting and Correcting Test Item Bias with a Logistic Response Model*. Statistical Laboratory, Department of Education, University of Chicago, Chicago, Illinois.

19   Guessing in Multiple Choice Tests

H. J. Rogers

Generally defined, guessing of test items occurs whenever an examinee responds to an item with less than perfect confidence in the answer. As such, guessing may be "blind", "cued" or "informed". Blind guessing occurs when an examinee has no idea of the correct answer and responds randomly. Cued guessing occurs on selection-type test items when an examinee does not know the answer and responds to some stimulus in the item such as an unintended clue or intentional mislead. Informed guessing occurs when the examinee's response is based on partial knowledge about the question.

Guessing of one form or another can occur on any type of test item, whether multiple choice or free response. However, it is generally considered to be a serious problem only on multiple choice tests, because of the greater likelihood of a correct answer through guessing. The primary psychometric problem arising from guessing on test items is that it increases the error variance of test scores, thereby reducing their reliability and validity. Since the 1920s, when multiple choice tests came into widespread use, there has been considerable research on ways to reduce the effects of guessing on test scores. This entry considers the many different approaches that have been proposed, reviews the research carried out to examine these approaches, and assesses the gains and losses associated with each approach.

1   Corrections for guessing

1.1   Formula scoring

Much of the research on corrections for guessing has focused on correction formulas applied to the scores after testing. The most widely used correction formula is based on the assumption that an examinee either knows the correct answer and chooses it, or does not know the answer and omits or responds randomly. From this perspective, wrong answers are the result of unlucky guessing; if this is so, the number of wrong answers can be used to predict the number of lucky guesses, which can then be deducted from the examinee's score. For example, on a four-option multiple choice test, a randomly guessing examinee will average one right answer (lucky guess) for every three wrong answers (unlucky guesses); thus, a third of the number of wrong answers should be deducted from the examinee's score.

The standard correction for guessing is given by the formula

$$F = R - \frac{W}{(A-1)}$$

where R is the number of right answers, W is the number of wrong answers, and A is the number of response options per item. Associated with the use of the correction formula are test directions that instruct examinees to omit items rather than guess randomly, as wrong answers will be penalized. Omits are not counted as wrong answers in this formulation.

A corresponding correction can be applied to the item difficulty index or p-value (the proportion of examinees answering the item correctly). The corrected p-value is given by the formula (Thorndike, 1982).

$$p_c = p - \frac{p_w}{(A-1)}$$

where p is the proportion of examinees answering the item correctly and p_w is the proportion of examinees attempting the item who answered it incorrectly. A problem with this correction is that when the proportion of correct answers falls below the chance level (due to particularly attractive distractors), it is possible to obtain a corrected item difficulty index less than zero (Thorndike, 1982).

Argument about the appropriateness and effectiveness of formula scoring and the associated test directions continued in the early 1990s. Lord (1975) remarked that formula scoring is an area "where two informed people often hold opposing views with great assurance" (p. 7). Critics of formula scoring claim that: (a) it is based on false assumptions about examinee behavior, and (b) it penalizes examinees with certain personality traits.

With respect to the first point, critics argue that examinees who do not know the answer to a test item rarely respond randomly. If this were so, the distractors for an item would be chosen with approximately equal frequency. This is rarely the case, as can easily be shown. Thorndike (1982) illustrated this point with the example of a set of verbal analogy items from a published test, where the most popular distractor was, on average, chosen by about 20 per cent of examinees and the least popular by about 4 per cent of examinees.

Another criticism of the formula-scoring assumption is that it ignores the possibility that an examinee has either misinformation or partial knowledge (Rowley and Traub, 1977). If examinees respond on the basis of misinformation, the formula overcorrects, since examinees lose not only the point for the question but an additional fraction of a point for behavior that cannot be considered to be guessing. If informed guessing based on partial knowledge is held to be undesirable, then the formula undercorrects, since examinees are more likely to get an item right in this case than if they had been randomly guessing.

Whether informed guessing should be penalized at all is another issue. It has been argued that informed guessing increases true score variance rather than error variance and hence enhances the validity of scores (Mehrens and Lehman, 1986). Moreover, when examinees respond on the basis of partial knowledge, their test scores are based on a greater sample of content than if they omit, as instructed under formula-scoring directions, and hence may have greater validity. If examinees with partial knowledge are not to be penalized, the test directions should instruct examinees to omit only if they cannot eliminate any of the answer choices.

In answer to criticisms of the formula-scoring assumption, Lord (1975) described an alternative assumption which provides a more defensible basis for formula scoring. Lord's assumption is that the difference between an answer sheet obtained under formula-scoring directions and the same answer sheet obtained under number-rights directions is that omits, if any, on the former, are replaced by random guesses on the latter. Under this assumption, it can be shown that while the number-right and formula scores are unbiased estimators of the same quantity, the formula score has smaller error variance and is therefore preferable. Rowley and Traub (1977) pointed out that Lord's assumption is tenable only if examinees are able to distinguish informed guesses from random guesses. Some studies (e.g., Sheriffs and Boomer, 1954; Slakter, 1968; Votaw, 1936) have shown, however, that under formula-scoring directions, examinees omit items which they have a greater-than-chance probability of answering correctly. Lord (1975) criticized these studies on methodological grounds.

With respect to the argument that formula-scoring confounds personality traits with test scores, there is a considerable body of research which shows that the extent to which examinees comply with the instructions associated with formula scoring (i.e., the instruction to omit rather than to guess randomly) reflects a personality trait which may disadvantage some examinees (see Diamond and Evans, 1973; Rowley and Traub, 1977, for reviews of this research). Examinees who are unwilling to take risks may omit rather than answer a question they feel uncertain about even if they have partial knowledge, and consequently lose more points than they would have by answering. Examinees who are more willing to take risks will not be penalized on average, since at most they will lose the points gained by randomly guessing. The research in this area indicates that the tendency to omit under formula-scoring directions is a personality trait which is more reliably measured by multiple choice tests than the cognitive trait of interest.

The primary argument in favor of formula scoring is that it increases the reliability and validity of scores (Mattson, 1965; Lord, 1975). While the argument has theoretical merit (under the assumptions of formula scoring), empirical studies have not yielded unequivocal support (Diamond and Evans, 1973; Rowley and Traub, 1977). Increases in reliability and validity, where obtained, are generally small (for a review, see Diamond and Evans, 1973). Rowley and Traub (1977) suggested that even when increases in reliability are obtained, the increase may reflect the reliability of measurement of the personality trait rather than a reduction in the error variance of the true scores on the cognitive component.

A second argument in favor of formula scoring is based on empirical studies that show that formula scoring tends to equalize the mean scores of randomly equivalent groups of examinees who have been given different instructions regarding guessing. Angoff and Schrader (1984) compared the mean number-right and formula scores of groups of examinees who were given either number-right or formula-scoring directions and found that while there was a significant difference in the mean number-right scores of the two groups, there was almost no difference in the mean formula scores. The authors concluded that the difference in number-right scores was due largely to chance responding. Angoff and Schrader (1986), using the same data, drew the additional conclusion that partial knowledge is often misinformation: that is, low-ability examinees with misinformation are deterred from responding under formula-scoring directions (and hence lose fewer points than they would have under number-right scoring) as often as higher-ability examinees with partial knowledge (who lose more points than they would have otherwise), with the result that there is no net effect on formula scores. This conclusion, however, does more to support the argument that the mean scores of the groups are an inappropriate basis for comparison of the effects of test directions than it does to support the argument in favor of formula scoring.

An alternative formula score used by Traub *et al.* (1969) is given by

$$F = R + \frac{O}{A}$$

where O is the number of omits. Under this procedure, examinees gain a chance score on items they omit, rather than a penalty for wrong answers. This formula scoring procedure may be less disadvantageous to the timid examinee (Mehrens and Lehman, 1986), but on the other hand, may disadvantage better students who have partial knowledge but who prefer to accept the omit reward rather than attempt the question (Wood, 1976). Hence, personality factors may once again be confounded with scores obtained under formula-scoring directions.

Lord (1975) noted that the omit-reward formula produces scores which are perfectly correlated with standard formula scores. This is true, however, only if the scores are obtained under the same testing conditions. The scores will not be perfectly correlated when each is obtained under directions appropriate to the procedure, since the directions will differentially affect omitting behavior.

1.2   Item response theory approaches

Item response theory (IRT) is a test theory which has come into wide use only since the early 1970s. Item response theory provides an alternative approach to the problem of guessing. Under IRT, an examinee's observed performance on a test item is assumed to depend on the examinee's underlying and unobservable ability or trait level and characteristics of the test item. The relationship is expressed in the form of a mathematical function, or item characteristic curve, which gives the probability of a correct response to an item as a function of the person and item parameters. Under item response theory, the examinee ability estimate is not a simple transformation of number-correct test score; it is estimated in the presence of parameters describing the test item, and hence takes into account the characteristics of the item (see *Item Response Theory*).

The most general of the commonly used item response models, referred to as the "three-parameter model", assumes that examinee performance on a test item is affected by three characteristics of the item: its difficulty, discrimination, and a factor which reflects the probability that a very low ability examinee will answer the item correctly. This parameter of the item is denoted the *c*-parameter and is sometimes referred to as a "pseudoguessing" parameter. Items differ in their *c*-parameters according to their difficulty and the attractiveness of the distractors. The three-parameter item response model is given by the equation

$$P(u=1\mid\theta)=c+(1-c)\frac{e^{a(\theta-b)}}{1+e^{a(\theta-b)}}$$

where *u* is the examinee's response to the item (taking the value 1 for a correct response), θ is the examinee's ability, *a* is the item discrimination parameter, *b* is the item difficulty parameter, and *c* is the pseudoguessing parameter. Examples of three-parameter item characteristic curves are shown in Figure 1.

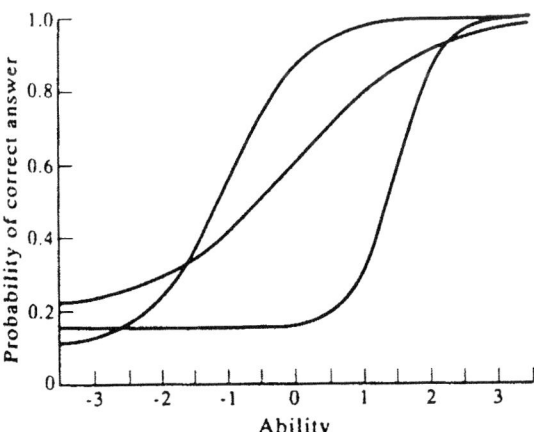

Figure 1   Three-parameter item characteristic curves.

While the c-parameter does indicate that there is a nonzero probability of a correct answer even for the lowest ability examinees, it does not indicate the probability of a correct answer through random guessing. The c-parameter for multiple choice items is often lower than the chance probability of a correct answer because of the attractiveness of the distractors.

Although the c-parameter is considered to be a characteristic of the item rather than the examinee, its effect does depend on the ability level of the examinee. Hence, it does take into account the individual examinees' differential probabilities of guessing and its incorporation in the model allows for the adjustment of the ability estimate to take into account the possibility of guessing. However, whenever the c-parameter is nonzero, the precision of estimation of ability is reduced. Hence, as in the case where test score is used as the trait estimate, the occurrrence of guessing increases error variance and is therefore undesirable.

In practice, satisfactory parameter estimation in the three-parameter model requires large samples of examinees and long tests. The c-parameter is often poorly estimated, even when data are abundant. For this reason, many practitioners choose more restrictive models, incorporating only one or two item parameters. These models are easier to fit to test data. However, one- and two-parameter models assume that very low ability examinees have zero probability of obtaining a correct answer, and therefore make no allowance for guessing behavior. If guessing is a factor in examinee performance, the ability of the examinee will be overestimated and the estimate will contain greater error.

Waller (1989) suggested a modification of the one- and two-parameter models which would yield an ability estimate corrected for guessing without including a lower asymptote parameter. Waller suggested that at each stage of the estimation process, an examinee's ability estimate be based only on the items for which the examinee has a greater-than-chance probability of answering correctly. Waller's initial evaluation of this model indicated that it does not fit real data as well as the three-parameter model. Waller concluded that the three-parameter model is better able to model partial-knowledge guessing, since the c-parameter has an effect throughout the ability range, whereas Waller's model only considers random guessing, which is most likely to occur among low ability examinees. There appears to have been no follow-up research with respect to this model.

1.3 Other approaches

Two other testing procedures designed to circumvent the problem of guessing in multiple choice tests and allow for assessment of partial knowledge are confidence testing and elimination testing procedures.

Confidence testing requires examinees to indicate their degree of confidence in each of the response options for an item or simply in the option they choose; guessing and partial knowledge can thus be identified. Various schemes for allotting confidence points have been proposed. Echternacht (1972) provided a review of confidence testing procedures.

Studies of confidence testing procedures have produced mixed results with respect to improvements in reliability and validity of scores. For example, Hambleton *et al.* (1970) found that confidence testing produced more valid but less reliable scores; Hakstian and Kansup (1975) concluded that confidence testing procedures do not result in any increase in reliability or validity. An obvious disadvantage of confidence testing procedures is that they require increased testing time and are more difficult for examinees. In addition, there is some evidence (Jacobs, 1971; Hansen, 1971) that the personality factor of "general confidence" contaminates scores obtained under confidence testing procedures.

Elimination testing requires examinees to eliminate the answer choices they believe to be incorrect, with points awarded for each incorrect option eliminated and points deducted if the correct answer is eliminated. There are many variants on the basic procedure. As with confidence testing procedures, research has shown no clear advantage to elimination procedures. Hakstian and Kansup (1975) summarized the research in this area and observed that there was no evidence of an increase in reliability and, at best, provisional evidence for an increase in validity. These authors concluded on the basis of their own study that there was no empirical support for elimination testing procedures.

Because of the lack of empirical evidence supporting the use of these procedures, neither is widely used and there has been little research in this area since the 1970s.

2   Detection of guessing in multiple choice tests

Apart from procedures for correcting for guessing in multiple choice tests, a number of procedures exist for detecting "aberrant" test-taking behavior such as guessing. These procedures are known generally as "appropriateness measures".

Perhaps the simplest appropriateness measure is the personal biserial correlation (Donlon and Fischer, 1968). The personal biserial correlation coefficient is a measure of the relationship between the examinee's response pattern and the item difficulties (p-values). When the examinee responds in a consistent manner, he or she should answer easier items correctly more often than difficult items, hence the correlation should be high; if the examinee is randomly guessing, then he or she will obtain a correct answer on more difficult items than expected, leading to a lower correlation coefficient.

Other appropriateness measures are based on a comparison of the examinee's observed response pattern with that expected under certain assumptions. One such measure is Sato's Caution Index (Sato, 1975), which is obtained by first computing the covariance between the item difficulties and the examinee response pattern; this covariance is then compared with the covariance between item difficulty and the perfect Guttman response pattern for an examinee with that total score. The ratio of these quantities will be close to one when the examinee responds in a consistent manner and close to zero when the response pattern is extremely aberrant.

A number of appropriateness measures are based on item response theory. An example of an appropriateness measure based on IRT is the person-fit statistic, which can take a variety of forms. A general person-fit statistic is the mean squared standardized residual, given by

$$f=\frac{1}{n}\sum_{i=1}^{n}\frac{[u_i-P(\theta)]^2}{P(\theta)[1-P(\theta)]}$$

where n is the number of items, u_i is the response of the examinee to item i, and $P(\theta)$ is the probability that an examinee with ability θ will answer the item correctly. The probability of a correct answer is obtained using the parameter estimates obtained under the chosen item response model.

Other non-IRT and IRT-based appropriateness measures are described and evaluated in Harnisch and Linn (1981) and Drasgow *et al.* (1987). These appropriateness measures are designed to detect aberrant response patterns, but not to correct test scores or ability estimates for the aberrant responses. If an examinee's appropriateness measure indicates an aberrant response pattern, the examinee must be retested to obtain a more valid and reliable estimate of ability or achievement.

3 Conclusion

Despite over 50 years of research, there appears to be no generally accepted and satisfactory way of correcting for guessing in multiple choice test scores. While arguments have been made for and against formula scoring, there is no conclusive empirical evidence that formula scores are more reliable and valid than number-right scores, but some evidence suggests that formula-scoring test directions lead to the confounding of a personality factor with test score. Item response theory offers an alternative approach, but the practical problems of implementing IRT procedures may preclude the use of such procedures in some testing situations. Indices for detecting aberrant response patterns are available, but these measures offer no aid in correcting for the guessing observed.

Practical steps that can be taken to ameliorate the problem of guessing include ensuring that power tests are not administered under speeded conditions, that directions about guessing and scoring are clearly stated, that the test is of sufficient length so that guessing cannot substantially influence scores, and that the test is of appropriate difficulty for the examinees being tested.

References

Angoff, W. H. and Schrader, W. B. 1984. A study of hypotheses basic to the use of rights and formula scores. *J. Educ. Meas.* **21**(1), 1–17.
Angoff, W. H. and Schrader, W. B. 1986. A rejoinder to Albanese, "The correction for guessing: A further analysis of Angoff and Schrader". *J. Educ. Meas.* **23**(3), 237–43.
Diamond, J. and Evans, W. 1973. The correction for guessing. *Rev. Educ. Res.* **43**(2), 181–91.
Donlon, T. F. and Fischer, F. E. 1968. An index of an individual's agreement with group determined item difficulties. *Educ. Psychol. Meas.* **28**, 105–13.

Drasgow, F., Levine, M. V. and McLaughlin, M. E. 1987. Detecting inappropriate test scores with optimal and practical appropriateness indices. *Appl. Psychol. Meas.* **11**(1), 59–79.

Echternacht, G. J. 1972. The use of confidence testing in objective tests. *Rev. Educ. Res.* **42**(2), 217–36.

Hakstian, A. R. and Kansup, W. 1975. A comparison of several methods of assessing partial knowledge in multiple-choice tests: II. Testing procedures. *J. Educ. Meas.* **12**(4), 231–9.

Hambleton, R. K., Roberts, D. M. and Traub, R. E. 1970. A comparison of the reliability and validity of two methods for assessing partial knowledge on a multiple-choice test. *J. Educ. Meas.* **7**(2), 75–82.

Hansen, R. 1971. The influence of variables other than knowledge on probabilistic tests. *J. Educ. Meas.* **8**(1), 9–14.

Harnisch, D. L. and Linn, R. L. 1981. Analysis of item response patterns: Questionable test data and dissimilar curriculum practices. *J. Educ. Meas.* **18**, 133–46.

Jacobs, S. S. 1971. Correlates of unwarranted confidence in responses to objective test items. *J. Educ. Meas.* **8**(1), 15–19.

Lord, F. M. 1975. Formula scoring and number-right scoring. *J. Educ. Meas.* **12**(1), 7–11.

Mattson, D. 1965. The effects of guessing on the standard error of measurement and the reliability of test scores. *Educ. Psychol. Meas.* **25**(3), 727–30.

Mehrens, W. A. and Lehman, I. I. 1986. *Using Standardized Tests in Education*, 4th edn. Longman, New York.

Rowley, G. L. and Traub, R. E. 1977. Formula scoring, number-right scoring and test taking strategy. *J. Educ. Meas.* **14**(1), 15–22.

Sato, T. 1975. *The Construction and Interpretation of S-P Tables*. Meiji Tosho, Tokyo.

Sheriffs, A. C. and Boomer, D. S. 1954. Who is penalized by the penalty for guessing? *J. Educ. Psychol.* **45**: 81–90.

Slakter, M. J. 1968. The effect of guessing strategy on objective test scores. *J. Educ. Meas.* **5**(3), 217–22.

Thorndike, R. L. 1982. *Applied Psychometrics*. Houghton Mifflin, Boston, Massachusetts.

Traub, R. E., Hambleton, R. K. and Singh, B. 1969. Effects of promised reward and threatened penalty on performance on a multiple-choice vocabulary test. *Educ. Psychol. Meas.* **29**(4), 847–61.

Votaw, D. F. 1936. The effect of do-not-guess directions upon the validity of true-false or multiple-choice tests. *J. Educ. Psychol.* **27**, 698–703.

Waller, M. I. 1989. Modeling guessing behavior: A comparison of two IRT models. *Appl. Psychol. Meas.* **13**(3), 233–43.

Wood, R. 1976. Inhibiting blind guessing: The effect of instruction. *J. Educ. Meas.* **13**(4), 297–308.

Further reading

Hambleton, R. K., Swaminathan, H. and Rogers, H. J. 1991. *Fundamentals of Item Response Theory*. Sage, Newbury Park, California.

20 Measurement of Judgments

J. M. Linacre

The use of judges to rate examinee performance on test items is often necessary, but can cause a severe measurement problem. The Spearman true score model solution assumes a linear rating scale used by judges to rate each examinee at a true score with some error. True scores are the resultant measures but they are sample-dependent numerations, linear by assertion, but not in construction. Inter-rater reliability, the usual statistic of judge agreement, is of doubtful value, particularly when the intention is to measure individual examinee performances. Nevertheless, when the empirical departures from linearity are small and some components underlying the observations can be held to be inter-changeable, generalizability theory – an elaboration of the true-score model – provides some useful information. In contrast, a Rasch model analysis capitalizes on the inevitable judge disagreement to construct a sample-free, objective and linear measurement continuum. Models for both rating scales and rank orderings can be implemented. When true-score models are used, then equating of judged-tests is, at best, norm-referenced and hence sample-dependent. When Rasch models are used, the equating is criterion-referenced and hence sample-free.

1   Judgment and measurement in education

Expert judgment is required to rate the level of performance in many fields of education: written composition, music, drama, spoken language. The purpose of judging is to determine the best performances in a competition or what level of competence has been reached in a skill. The tasks, skills, or behaviors on which the examinees are rated will be called "items". All judges may rate all examinees on all items, or individual judges may be assigned to rate examinees on different items according to some judging plan. In practice, the optimum judging plan minimizes the number of judgments necessary to measure performances with the required precision.

Judgment is usually recorded by a judge making an independent ordinal rating of an examinee's performance on an item. The numeration of the discrete ratings on the scale is chosen for convenience in recording and ordering the categories. The situation may be as simple as each judge rating each examinee with 1 for success or 0 for failure on the only item, or as complex as each examinee performance involving several items, each with its own rating scale. The possible responses on the rating scales are categories defined in ascending order in terms of the performance level they represent (see *Rating Scale Analysis*).

The aim of the analysis is to derive from the ordinal, discrete ratings given by the judges final linear measures of performance for each examinee that are as fair and accurate as possible. This means that the measure given to an examinee must be estimated from, but be made statistically independent of, the particular judge or judges who rated that examinee. That is, the measure must be judge-free. In many situations, the measure must also be item-free; that is, independent of the particular items on which ratings were made. If the measure is also made independent of consideration of other examinees (i.e., person-free or sample-free) the measure can be termed "objective". Measures are usually subjected to further statistical analysis, such as means and standard deviations, that require measure linearity, not merely ordinal numeration over a finite interval, for meaningful interpretation (Stevens, 1959).

When judging is repeated at a later time, but standards are to be maintained, or examinee performances are to be compared across sessions, there is also the requirement that examinee measures be independent not only of the composition of the judging panel, and of the particular items chosen for recording competence, but also of the local levels of performance of the examinees at any session. Consequently, the most effective measurement model is not one that most completely describes any particular judging situation, but rather one that yields the most stable basis for inference – that is, the most objective measures.

2   Measuring by true scores

The perfect judging situation is often proposed to be that in which the ideal judge rates each examinee performance exactly at its "true score". The ratings given by empirical judges, however, rarely show complete agreement, and are therefore regarded as combining this fictitious "true score" with consistent judge differences and random error. This is analogous to the true-score model for objective testing and suffers from all its shortcomings.

Studies of judge behavior have identified numerous reasons for consistent differences between judges' ratings: leniency (severity, level, halo effect and central tendency (Guilford, 1954). Consistent differences are also found in interactions of judges with other aspects of the situation. A judge may be consistently lenient on one item, but consistently severe on another. Whatever part of each rating cannot be accounted for as a sum of the factors specified in the true-score model is treated as judge-generated random error (see *Rating Scale Analysis*).

Numerous studies have attempted to recover examinees' true scores from empirically observed ratings. The means of performing this is to construct a linear model in the rating metric of the form (Saal *et al*., 1980).

$$\text{Observed Rating} = \text{True score} + \text{Judge effect} + \text{Interaction} + \text{Error} \qquad (1)$$

In spite of the effort expended on this model, the resultant true scores have proved deficient. Specifications of rating scales have been arbitrary and, despite the numerical labeling, nonlinear in function. Thus, on a rating scale labeled from 1 to 5, the difference in performance between 1 and 2 is unlikely to be the same as that between 2 and 3. Consequently, an estimated true score of 1.7 has no clear quantitative meaning.

For most applications of these models, true scores are only estimable when every judge rates every examinee on every item or when a sophisticated judging plan, such as a balanced incomplete block design (Braun, 1988), is followed scrupulously. These constraints are impractical for large-scale test administration and so are seldom used.

The introduction into the measurement model of interaction effects, such as each judge's personal degree of bias against individual examinees, further limits the objectivity of the resultant measures. This is because interactions are local to the specific judging session. Better judge training is often proposed as the way to reduce interaction effects and error variance. Though judge training is beneficial, there is no example in the literature where it has eliminated judge differences.

3   Inter-rater reliability

An inter-rater (inter-judge) reliability coefficient, analogous to the reliability coefficient of objective tests, is often reported as a measure of judge agreement. This coefficient ranges from 0 to 1 with higher values regarded as better. Even though the reliability coefficient is manifestly insensitive to differences in judge severities, a high value is treated as evidence that the sums of examinee ratings accurately reflect examinee performance.

In general, the greater the ratio of true score variance to error variance in Equation (1), the higher the reliability. This reliability is dominated by the range of examinee ability, the range of judge severity, and the range of item difficulty. Since these ranges are local to the judging situation, any particular reliability lacks general significance. Some research even reports the paradoxical result that "examiners who agree are likely to be wrong" (Harper and Misra, 1976, p. 260).

4   Generalizability theory

Generalizability theory, an elaboration of true-score theory, also mistakes ratings as linear. It specifies judging designs which, with additional assumptions, are intended to lead to estimates of measurement error. An implementation of this approach is the GENOVA computer program (Crick and Brennan, 1982).

In a typical study, ratings are awarded by judges to examinees on items of performance according to a judging plan. The analyst specifies which aspect (examinees, judges, or items of performance) is the one to be measured. The other aspects, called "facets", are then sources of error. Error variances are calculated according to the judging plan and whether each facet is a random sample from a normally distributed admissible universe (e.g., of similar judges) or is the entire universe (e.g., all of a standardized test). These error variances are used to estimate a measurement error for the measure of interest.

If further, similar ratings are to be collected, a decision study can be performed to predict the error variances produced by revised judging plans. In particular, the decision study predicts what alterations to the previous judging plan would most reduce measurement error.

Generalizability theory is more successful when judges are in close numerical agreement, the test items are of similar difficulty, and the ratings are around the central, close to linear, part of the ordinal rating scale. In practice, the success of a generalizability study is actually evaluated only by the size of the error variances. Fit analysis and investigation of unexpected residuals is not undertaken.

5 Many-facet Rasch measurement

Many-facet Rasch models address the inevitable error in the ordinal ratings by specifying that each examinee has an estimable probability of being awarded any possible rating on any item by any judge (see *Rasch Measurement Theory*). The data are evaluated as potential realizations of these probabilities. Each observation is modeled as an ordinal rating stochastically governed by a log–linear combination of the measures of one element from each facet (e.g., an examinee ability or performance from the examinees' facet, a judge severity from the judges' facet, and an item difficulty from the items' facet). These measures are estimated so that they are statistically independent of the particular observations comprising the data, but they are realized through the data. The measures estimated for the elements maximize the likelihood of the observed ratings. An asymptotic standard error for each measure is provided. An implementation of this model is the *Facets* computer program (Linacre, 1988).

These measures constitute a linear measurements system that locates all elements of all facets on a common interval continuum of infinite range. The item measures on this continuum define the underlying variable (see *Item Response Theory*). The depiction of all measures as points along a line has the familiar visual interpretation of equal distances that is so productive in physical science and engineering.

The residual differences between the data and their measure-based expected values become the basis of detailed quality-control fit statistics and rating diagnoses. Significant misfit associated with particular elements and unexpectedly large residuals for particular observations identifies and diagnoses anomalies. Remedial action can be taken and errant (highly improbable) observations can be evaluated for re-rating or omission prrior to reporting.

5.1 Rasch judging plans

The only constraint on the judging plan is that the ratings must form a network of connections such that separate but comparable measures can be obtained for each examinee, judge and item. This does not require that every examinee be rated by every judge or even that any single observation be replicated, that is, rated more than once. The requirement is only that a chain of observations connects every judge, item, and examinee directly or indirectly. This has proved usefully robust in practice even with idiosyncratic judging plans. Since only recorded observations are used to estimate measures, missing data or even mistakes in implementing the judging plan do not threaten the analysis or require remedial maneuvers. Missing data have no practical consequence so long as the available ratings form one connected network of the type discussed by Luce and Perry (1949).

Judging plans have succeeded in which each examinee is rated by only one judge on any particular item. The judges rotate across items and examinees so that every examinee is rated by more than one judge, although on different items. Each judge rates more than one item, but for different examinees (Lunz et al., 1990). There is considerable flexibility because there is no requirement that each judge rate the same number of examinees or use the same number of items.

Defective plans are those in which judges are specialists who rate only items in their own specialties. Then it becomes impossible to determine how much of the challenge to the examinee is due to the difficulty of the specialty and how much is due to the severity of the specialty judges. Other defective plans are those in which one panel of judges rates one group of examinees, while another independent panel rates another independent group. Then it is impossible to determine whether mean differences between the groups are due to a mean ability difference among examinees or a mean severity difference among judges. Even these defective plans can be used if reasonable assertions beyond the data can be made, such as the equal severity of randomly assigned judging panels or the equal ability of randomly sampled examinee groups.

5.2 Dichotomously judged items

Dichotomous items produce observations in one of two categories, usually labeled 0 and 1, and generally coded so that 1 means success on the item and 0 means failure. A relevant many-facet Rasch model is:

$$\log(P_{nij1}/P_{nij0}) = B_n - D_i - C_j \tag{2}$$

where

P_{nij1} = the probability that person n will earn a 1 on item i from judge j;

B_n = ability of person n;
D_i = difficulty of item i;
C_j = severity of judge j.

Person abilities are not expressed in nonlinear raw scores on the particular test, but rather in linear log–odds units (logits) that measure examinees' levels of success on whatever items they performed, but in a general measurement system. The other facets are also calibrated in logits on the same continuum.

Solving Equation (2) for P_{nij1}, since $P_{nij0} = 1 - P_{nij1}$,

$$P_{nij1} = \frac{\exp(B_n - D_i - C_j)}{1 + \exp(B_n - D_i - C_j)} \tag{3}$$

brings out the exponential form of the model. If an examinee of ability 1 logit is rated by a judge of severity 0 logits on a dichotomous item of difficulty 0 logits, the probability of the rating being a 1 is $\exp(1)/(1 + \exp(1)) = 0.73$.

5.3   *Rating scales with discrete categories*

When examinees are rated by judges on several items, each of which has its own ordinal rating scale, a relevant many-facet Rasch model is:

$$\log(P_{nijk}/P_{nijk-1}) = B_n - D_i - C_j - F_{ik} \quad k = 1, K_i \tag{4}$$

where

P_{nijk} = the probability that person n will earn a rating of category k on item i from judge j;

B_n, D_i, C_j are defined as above;

F_{ik} = additional difficulty of being rated in category k beyond that of being rated in category $k - 1$ for item i;

K_i = the highest rating scale category, after ordinally renumbering of categories from 0.

The empirically observed rating scale is analyzed as ordinal categories representing increasing increments of performance. This redefinition replaces the arbitrary numeration of the original scale with a scale labeled in successive integers from 0 to K_i, in which each category label becomes the count of observed increments in performance above the lowest observed performance level.

Item difficulty, D_i, is usually defined such that the sum of all F_{ik} is set at 0. Thus the item difficulty, D_i, is calibrated at that point on the logit continuum at which the highest and lowest rating categories have equal probabilities of being awarded.

Zero or perfect scores are empirically inestimable. If there is a relevant context in which they can be embedded, they can be forced into the estimation procedure by means of simple Bayesian techniques.

5.4 *Rating scales with unobserved categories*

The many-facet model for discrete intervals renumbers the observed ratings into an ordinally ascending rating scale. Each category of this new rating scale represents a count of increments of performance and has associated with it a difficulty measure. In practice, not all categories may be observed. When the unobserved categories are an artifact of sampling, a simple computational device suffices to maintain estimability (Wilson, 1991). If, however, categories are unobserved because there are too many (e.g., a 0–100 rating scale), then representing the rating scale by a continuous rather than a discrete function can be useful.

A model for a continuous 0–100 rating scale, parallel to the discrete example above, is:

$$P_{nijk} = \frac{\exp(k(B_n - D_i - C_j) - G_{ik})}{\displaystyle\int_{m=0}^{100} (\exp(m(B_n - D_i - C_j) - G_{im})\, \mathrm{d}_m)} \tag{5}$$

where

P_{nijk} = the probability density that person n will earn a rating in category k on item i from judge j;

B_n, D_i, C_j are defined as above;

G_{ik} is a continuous function of k defining the rating scale structure for item i such that $G_{i0} = G_{i100} = 0$.

The function, G_{ik}, models the observed and unobserved parts of the rating scale. Since it is incompletely observed in the empirical data, its form must be specified by the analyst. Since the aim of the measurement model is to assist in interpretation and inference, it follows that the useful functions, G_{ik}, are ones with a few, readily interpretable parameters, such as polynomial or Fourier series (Andrich, 1982; Muller, 1987).

6 Measuring by rank ordering

Difficulties experienced in defining and using rating scales have led some analysts to prefer ranking examinees. Even when competent judges cannot agree on the category into which to place a performance, they may agree on which performance is better. This approach is attractive when the purpose is to find the best of a small number of examinees. When there is a criterion performance standard, this criterion performance can be ranked along with examinees to identify a pass–fail point.

The extent of agreement among judges ranking examinees depends on the variance in examinee performance. If several examinees have similar abilities, there can be little agreement among judges' rank-orderings (Harper and Misra, 1976). Thus the reliability of rank-ordering is also sample-dependent. Nevertheless, summing rankings is often satisfactory for the purpose at hand.

6.1   Rasch models for rank ordering

Lack of agreement between rank-orderings from different judges can be evaluated with a Rasch measurement model, and so used to obtain linear measures of the distances between examinees. The model for the simplest case, paired comparisons, is:

$$\log(P_{nmij}/(1 - P_{nmij})) = B_n - B_m \tag{6}$$

where

P_{nmij} is the probability that person n is preferred to person m on item i by judge j;
B_n = ability of person n;
B_m = ability of person m.

This model enables measures, standard errors, and fit statistics to be obtained for each person. Though the judges and items are not parameterized explicitly, data partitioning enables quality-control fit statistics to be obtained for each judge and item. Thus idiosyncratic judging and uneven examinee performance across items can be detected.

Rank-ordering of more than two examinees introduces interexaminee observation dependency not present in ratings. Though this can be modeled explicitly, empirical studies indicate that modeling the rankings as ratings provides a practical approximation. Such a model, allowing tied and partial rankings, is:

$$\log(P_{nijk}/P_{nijk+1}) = B_n - F_{ijk} \qquad k = 1, K_{ij} - 1 \tag{7}$$

where

P_{nijk} = the probability that person n will be ranked k on item i by judge j;
B_n is defined as above;
F_{ijk} = additional difficulty of being ranked k beyond that of being ranked lower at $k + 1$ in the rank-ordering on item i by judge j;
K_{ij} = the number of different rankings made by judge j on item i. The highest ranking is 1.

7   The equating of judged tests

Equating judges from one testing session to another using true-score models remains intractable. Generalizability theory provides a mathematical solution provided that no variation in overall judge behavior or item or judge sampling occurs. Many examination boards attempt a solution by assuming that their distributions of examinee performances are the same from testing session to testing session. They then apply an arbitrary norm-referenced rule, say, that 80 percent of examinees shall always be said to succeed, whatever the actual level of competence.

Rasch measurement models yield individual judge severity estimates and individual item difficulty estimates. When the data fit the measurement model, these measures are statistically sample-free for judge, item, and examinee populations similar to those used to estimate these measures. Consequently, when there has been no substantive change in the corresponding judge or item, each measure can be used as maintaining its numerical value. The relative stability of measures can be verified by comparing the measures of common elements across testing sessions.

The characteristics of judges do change between, and even within, judging sessions. The characteristics of items change more slowly. Research indicates, however, that panels of experienced judges can maintain their overall group level of severity over time. Consequently, so long as each facet except one (usually examinees) of each new test has some components in common with the corresponding facets of the old test, the Rasch continua constructed for the two tests can be aligned; that is, equated. This enables the maintenance of criterion levels of performance (such as pass–fail points) from judging session to judging session despite the introduction of new judges and new items and the inevitable variations in distributions of examinee ability on the different occasions.

8   Unidimensionality

Both true-score and Rasch models specify that the variable to be measured is unidimensional. General lack of fit of data to model is an indication that the facets of the model do not add up to realize one variable in which greater competence by the examinees is expressed in terms of higher ratings awarded by the judges on the items. Such lack of fit is not unusual in practical examinations in which conflicting requirements such as speed, accuracy, and neatness are confounded. Under these circumstances, any decision based on a cumulative score involves an arbitrary weighting of the various non-homogeneous components. Unlike truescore models, which tend to hide the underlying ambiguity, Rasch fit statistics point it out to the examination board, on whom policy decisions concerning the relative importance of the different components depend (see also: *Rating Scale Analysis; Partial Credit Model; Rasch Measurement Theory; Rasch Measurement Models*).

References

Andrich, D. 1982. An extension of the Rasch model for ratings provided both location and dispersion parameters. *Psychometri.* **47**(1), 105–13.

Braun, H. I. 1988. Understanding scoring reliability: Experiments in calibrating essay readers. *J. Ed. Stat.* **13**(1), 1–18.

Crick, J. E. and Brennan, R. L. 1982. GENOVA: Generalized Analysis of Variance. Computer program. University of Massachusetts at Boston Computer Facilities, Dorchester, Massachusetts.

Guilford, J. P. 1954. *Psychometric Methods*, 2nd edn. McGraw-Hill, London.

Harper, A. E. and Misra, V. S. 1976. *Research on Examinations in India*. National Council of Educational Research and Training, New Dehli.

Linacre, J. M. 1988. Facets: Many-facet Rasch Measurement. Computer program. MESA Press, Chicago, Illinois.

Luce, R. D. and Perry, A. D. 1949. A method of matrix analysis of group structure. *Psychometri.* **14**(1), 95–116.

Lunz, M. E., Wright, B. D. and Linacre, J. M. 1990. Measuring the impact of judge severity on examination scores. *Applied Measurement in Education* **3**(4), 331–45.

Muller, H. 1987. A Rasch model for continuous ratings. *Psycxhometri.* **52**(2), 165–81.

Saal, F. E., Downey, R. G. and Lahey, M. A. 1980. Rating the ratings: Assessing the Psychometric quality of rating data. *Psych. Bull.* **88**(2), 413–28.

Stevens, S. S. 1959. Measurement, psychophysics and utility. In: Churchman, C. W. and Ratoosh, P. (eds.) 1959. *Measurement: Definitions and Theories*. Wiley, New York.

Wilson, M. 1991. Unobserved categories. *Rasch Measurement* **5**(1), 128.

Further reading

Brennan, R. L. 1992. *Elements of Generalizability Theory* Rev. edn. American College Testing Program, Iowa City, Iowa.

Cardinet, J. and Tourneur, Y. 1985. *Assurer la mesure*. Lang, New York.

Engelhard, G. 1992. The measurement of writing ability with a many-faceted Rasch model. *Applied Measurement in Education* **5**(3), 171–91.

Engelhard, G. and Wilson, M. (eds.) 1996. The Many Facet Rasch (Facets) Model. In *Objective Measurement: Theory Into Practice*. Vol. 3. Ablex, Norwood, New Jersey.

Heller, J., Sheingold, K., Nunez, A. and Myford, C. 1996. Examining the Quality of Rater Reasoning: Confronting Barriers to Valid and Reliable Portfolio Assessment. Educational Testing Service, Princeton, New Jersey.

Linacre, J. M. 1989. *Many-facet Rasch Measurement*. MESA Press, Chicago, Illinois.

Shavelson, R. J. and Webb, N. M. 1991. *Generalizability Theory*. Sage, Newbury Park, California.

21 Charting of Student Progress

G. N. Masters, R. J. Adams and M. Wilson

A common approach to measuring a student's achievement is to administer a test and to record the questions he or she answers correctly. In modern test theory (see *Rasch Measurement Theory ; Item Response Theory*), the resulting record of right and wrong answers leads to an estimate of that student's standing on an achievement continuum. This continuum is marked out and defined operationally by the test questions that students attempt and so provides a framework for charting progress over time.

This methodology is particularly appropriate if the aim of an instructional program is to provide students with a body of facts, skills and procedures. Test questions can be written to assess the presence or absence of specific items of learning on any given occasion and students' performances on those items can be scored either right or wrong. But not all learning is a matter of absorbing and reproducing provided information. An important form of learning occurs when students change the way they conceptualize aspects of a subject. Progress occurs when a student discards a naive model or representation of a phenomenon in favor of a more expert conception. Traditional achievement tests are not well suited to the identification of the conceptions that students bring to problems. To chart progress in conceptual understanding, an alternative testing methodology is required. This entry describes recent advances in developing such a methodology.

1 Learning as an active process

Implicit in much of our current measurement theory and practice is a view of learners as passive absorbers of provided wisdom. Most items on standard achievement tests assess students' abilities to recall and apply facts and routines presented during instruction. Some require only the memorization of detail; they seek evidence that students have absorbed factual details presented in class and are able to reproduce these on command. Other achievement items, although supposed to assess higher-level learning outcomes like "comprehension" and "application", often require little more than the ability to recall a formula and to make appropriate substitutions to arrive at a correct answer.

Test items of this type are consistent with a view of learning as a passive, receptive process through which new facts and skills are added to a learner's repertoire in much the same way as bricks might progressively be added to a wall. The process is additive and incremental: students with the highest levels of achievement in an area are those who have absorbed and can reproduce the greatest numbers of facts and formulae. The practice of scoring answers to items of this type either "right" or "wrong" is consistent with the view that individual units of knowledge or skill are either present or absent in a learner at the time of testing. Under this approach, diagnosis is a simple matter of identifying unexpected holes or gaps in a student's store of knowledge. These are subareas of learning in which knowledge is "missing" and in which there is a need for remedial teaching to fill a deficit.

This approach to the measurement of achievement may be appropriate for some forms of learning. However, in recent decades, significant advances have occurred in our understanding of the ways in which students learn. In particular, there has been an increased awareness of the active, constructive nature of most forms of human learning and of the important role that students' personal conceptions and representations of subject matter play in the learning process. Rather than being a passive process of absorbing new material as it is encountered, meaningful learning is increasingly being recognized as an active process through which students construct their own inter-pretations, approaches and ways of viewing phenomena, and through which learners relate new information to their existing knowledge and understandings. Under this view of learning, the difference between beginning and advanced learners is seen not so much as a difference in amount of factual knowledge (although this is usually an important aspect of competent performance), but as a difference in the types of conceptions and understandings that students bring to a problem, and in the strategies and approaches that they use.

Support for this view of learning can be found in recent studies in a number of areas of investigation. In cognitive science, comparisons of novices and experts in various fields of learning (Chi et al., 1981; Larkin, 1983; McCloskey et al., 1980) show that expertise typically involves much more than mastery of a body of facts: experts and novices usually have very different ways of viewing phenomena and of representing and approaching problems in a field. Expert–novice studies suggest that the performances of beginning learners often can be understood in terms of the inappropriate or inefficient models that these learners have constructed for themselves.

Similar observations have been made in the field of science education (see Driver and Easley, 1978; Osborne and Wittrock, 1983; Posner et al., 1982). Research into students' science learning has drawn attention to the frequent mismatch between intuitive understandings that students bring to the classroom and the conceptual frameworks assumed by teachers. Caramazza et al. (1981) observed that the scientific "principles" that students abstract from everyday experience are often strikingly at variance with the most fundamental physical laws. These misunderstandings can go undetected by teachers if correct answers to test questions depend only on superficial knowledge of formulas and formula manipulation techniques (Clement, 1982). There is evidence that students can succeed in high school and even college science courses while still maintaining many of their misconceptions and without acquiring an understanding of underlying principles (White and Horwitz, 1987).

Related work in Sweden (Marton, 1981; Dahlgren, 1984; Säljö, 1984) has used clinical interviews to explore the different understandings that students have of key principles and phenomena in a number of fields of learning. These interviews have revealed a range of student conceptions of each of the phenomena that these studies have explored and have illustrated the importance of forms of learning which produce "a qualitative change in a person's conception of a phenomenon" from a lower-level, more naive conception to a more expert understanding of that phenomenon (Johansson *et al.*, 1985, p. 235).

Under this view of learning, a student is rarely considered to have no understanding or no strategy when addressing a problem. Even beginning learners are considered to be engaged in an active search for meaning, constructing and using naive representations or models of subject matter. Rather than being "wrong", these representations frequently display partial understanding and are applied rationally and consistently by the individuals who use them. In arithmetic, for example, "it has been demonstrated repeatedly that novices who make mistakes do not make them at random, but rather operate in terms of meaning systems that they hold at a given time" (Nesher, 1986).

An implication of this view of learning for the assessment and monitoring of student learning is that much greater cognizance must be taken of the understandings and models that individual students construct for themselves during the learning process. In many areas of learning, levels of achievement might be better defined and measured not in terms of the number of facts and procedures that a student can reproduce, but in terms of his or her levels of understanding of key concepts and principles underlying a learning area.

2   Conventional achievement testing

Techniques for constructing achievement tests have been developed and refined over many decades. Most achievement tests begin with a statement of the instructional objectives to be assessed. According to Bloom *et al.* (1971, p. 28), these objectives should be stated as directly observable student behaviors which can be reliably recorded as either present or absent. They should be "stated in terms which are operational, involving reliable observation and allowing no leeway in interpretation". To achieve this degree of reliability, test constructors are encouraged to write items to assess students' abilities to perform unambiguous, observable tasks like "stating", "listing", "naming", "selecting", "recognising", "matching" and "calculating" (Bloom *et al.*, 1971, p. 34).

This emphasis on specifying and testing precise student behaviors has led to the construction of achievement tests composed of discrete items, each relating to a particular behavioral objective, and each scorable as either right or wrong. Multiple choice items have become especially popular in achievement tests because they can be scored quickly, unambiguously, and even by machine. In some areas of education, machine-scored multiple choice tests are now the principal mode of evaluating student learning.

The advantages of traditional achievement testing include the fact that it provides a close link between curriculum objectives which can be expressed in behavioral terms and the resulting measures of student achievement. By constructing standard conditions and establishing unambiguous right/wrong scoring rules, standardized achievement tests reduce subjectivity in assessment and provide results which are comparable over time and across students.

A potential disadvantage of conventional achievement tests is that, because of the emphasis these tests place on precisely-defined student behaviors, they can encourage students to focus their efforts on relatively superficial forms of learning. An alternative approach is to base achievement testing not on the detailed specification of many observable student behaviors, each of which can be recorded as either present or absent, but on a consideration of the key concepts, principles and phenomena that underlie a course of instruction and around which factual learning can be organized. Rather than recording students' understandings of these concepts as either "right" or "wrong", this alternative approach recognizes that learners have a variety of understandings of phenomena, and that some of these understandings are less complete than others.

3   Testing for conceptual understanding

An achievement testing methodology based on an active, constructive view of learning differs from more traditional approaches in that its prime focus is on revealing how individual students view and think about key concepts in a subject. Rather than comparing students' responses with a "correct" answer to a question so that each response can be scored right or wrong, the emphasis is on understanding the variety of responses that students make to a question and inferring from those responses students' levels of conceptual understanding.

One area of learning in which a great deal of work has been carried out to understand how students think about and approach phenomena is the area of physics education. Studies in a number of countries have explored students' understandings of such concepts as acceleration (Trowbridge and McDermott, 1981); mole concept (Lybeck et al., 1988); electric charge, enthalpy and entropy, force and motion (Viennot, 1979); gravitation (Champagne et al., 1980; Gunstone and White, 1981); light and the transmission of heat, momentum, potential difference, proportionality, torque and such principles and models as Newton's laws, conservation laws, the atomic model, and electron flow models for circuits.

A common technique in these studies is to use drawings of simple physical systems and to ask students to describe what is happening in each drawing (e.g., to predict what will happen to an object, to describe the forces acting on a body, or to draw the trajectory that an object will follow). During these interviews, students are asked to explain their responses and their explanations are tape recorded (Johansson *et al*., 1985; McCloskey, 1983).

In other studies, students are asked to manipulate apparatus in a laboratory to achieve particular effects (e.g., to apply a force to make a body move in a particular direction) and students' explanations and comments have been tape recorded and transcribed (McDermott, 1984). Still other researchers (e.g., diSessa, 1982; White, 1983) have developed interactive software for this purpose. In these studies, students are asked to apply "forces" to simulated objects on a screen to make them move to specified positions, to speed up, to slow down, and so on.

The observations made in these studies suggest that students do not simply make random "errors" but operate in terms of naive theories about physical phenomena. In the area of force and motion, these theories can be "remarkably well articulated, ... quite consistent across individuals, . . . and strikingly inconsistent with the fundamental principles of classical mechanics" (McCloskey, 1983, p. 299). In his studies of students' attempts to control a simulated object on a screen, diSessa (1982, p. 38) found "a surprising structure of discrete and definite theories" about how forces influence motion; and, through their interviews with Swedish students about aspects of science learning, Johansson *et al*. (1985) arrive at a similar conclusion.

> In our case, a discovery of decisive importance was that for each phenomenon, principle, or aspect of reality, the understanding of which we studied, there seemed to exist a limited number of qualitatively different conceptions of that phenomenon, principle, or aspect of reality. (Johansson *et al*., pp. 235–6)

A number of researchers have observed that the same naive conceptions can be found among students of different ages and with different educational backgrounds. McCloskey (1983), for example, found the same types of naive physical theories among students who had never taken physics, among high school physics students, and among college physics students. The only difference was in the frequencies of occurrence of these different understandings. McDermott (1984) reports an identical observation in a Norwegian study of high school physics students, future high school science teachers, and physics graduates.

A significant finding of these studies is that some students can succeed on traditional achievement tests and graduate from high school and even college physics courses with their naive conceptions of physical principles largely unchanged. Through their physics courses students are able to "master certain methods of calculation without having adopted the conceptualization underlying them" (Johansson *et al*., 1985, p. 235). Indeed, a misconception "may go undetected because a student's superficial knowledge of formulas and formula manipulation techniques can mask his or her misunderstanding of an underlying concept" (Clement, 1982, p. 66). The result is that "many students emerge from their study of physics and physical science without a functional understanding of some elementary but fundamental concepts" (McDermott, 1984, p. 31).

These findings invite a reconsideration of the way in which we think about and attempt to measure student learning. Clearly, many students are succeeding on precise, operationally-defined objectives without developing an understanding of the material that they are learning. For many science educators, the answer is to place greater emphasis not on the learning of scientific facts and formulas, but on changing students' ways of thinking about scientific phenomena:

> The formal learning of science can be viewed as involving, at least in part, a shift from one set of beliefs about the physical world to another, one set of conceptions to another. (Osborne and Wittrock, 1983, p. 81)

and

> In our view, learning (or the kind of learning we are primarily interested in) is a qualitative change in a person's conception of a certain phenomenon or of a certain aspect of reality. (Johansson et al., 1985, p. 235)

4   Constructing ordered outcome categories

A methodology for charting progress in conceptual understanding first identifies a range of key concepts in an area of learning and then develops questions or tasks that can be used to explore the different understandings that students have of those concepts. For each of these questions, a set of ordered categories is defined corresponding to different levels of conceptual understanding. This notion of order is basic to a view of learning as a "shift" or "change" in a student's understanding. Such a change constitutes "learning" only if it involves a change from a lower level, more naive understanding to a higher level, more expert conception.

In practice, the set of ordered categories for a question is constructed by first exploring the variety of responses that students give when confronted with that question and asked to explain their thinking about it. In other words, the data from which ordered categories are constructed for a question are usually collected through student interviews. The resulting categories provide a framework for recording future responses to that question and introduce the possibility of basing measures of achievement on inferences about students' levels of understanding.

The construction of categories of response from interview data is part of the method used by Marton (1981) and his colleagues at the University of Gothenburg. These researchers interview students to explore their understandings of particular concepts and principles, transcribe tape-recordings of these interviews and then carry out detailed analyses of transcripts. "The aim of the analysis is to yield descriptive categories representing qualitatively distinct conceptions of a phenomenon". These categories form an "outcome space" which provides "a kind of analytic map" of students' understandings of each phenomenon. Learning is thought of as "a shift from one conception to another" on this map (Dahlgren, 1984, p. 2431).

5   Collecting observations

While interviews with students are essential for identifying the variety of understandings that learners have of phenomena and for constructing ordered categories for individual questions, interviews are not practicable as a basis for achievement testing. For the purposes of assigning students to the categories that have been defined for test questions, alternative observation formats must be used. In general, this requires imaginative new kinds of tests capable of providing information about the conceptions that students bring to questions.

One approach to exploring students' levels of understanding is through computer-administered tasks. When students enter their responses to questions into a computer, these can be matched to libraries of common responses. In this way, particular kinds of errors and misunderstandings can be identified and inferences made about students' levels of understanding.

Alternatively, a student's initial response might trigger additional hints and subques-tions designed to provide further information about the nature of that student's understanding. Through additional questioning of this kind it may be possible to emulate in a crude way the type of exploration that can be carried out through an interview to reveal a student's conception of a phenomenon.

Achievement tests constructed to reveal students' models and representations of phenomena may bear little resemblance to traditional achievement tests. As diSessa (1982) and White (1983) show, a great deal of information can be collected about individuals' naive theories of force and motion by asking them to move simulated objects on a screen. A computer can be used to keep detailed records of when students apply forces, in which directions they apply those forces, and how they respond to the motion that they produce. Automatic analyses of student records can then be used to infer students' understandings of the relations between force and motion.

6   Constructing measures of achievement

The construction of achievement measures from observations recorded in sets of ordered outcome categories requires a psychometric method analogous to the methods that have been developed for the analysis of right and wrong answers to test questions, but designed for ordered response categories. Such a method is described by Masters (1982) and Wright and Masters (1982). This psychometric method (see *Partial Credit Model*) proposes that the probability of a person scoring x rather than $x - 1$ on a particular item i will increase steadily with ability in an area of learning such that

$$\frac{\pi_{nix}}{\pi_{nix} - 1 + \pi_{nix}} = \frac{\exp(\beta_n - \delta_{ix})}{1 + \exp(\beta_n - \delta_{ix})} \tag{1}$$

where π_{nix} is the probability of person n responding in category x ($x = 1, 2, \ldots, m_i$) of item i, β_n is person n's level of ability in this area of learning (to be measured by this set of items), and δ_{ix} is a parameter that governs the probability of a response being made in category x rather than in category $x - 1$ of item i.

The consequence of applying the simple logistic expression (1) to the transition between each pair of adjacent outcome categories for each item, is that a connection is formed between the ordered categories for that item and the underlying variable that the set of items is used to measure. It is this connection that enables performances on each item to be used to estimate students' locations on the underlying variable. This model permits measures of achievement to be constructed from inferences of students' levels of understanding of each of a number of concepts or phenomena in an area of learning.

6.1 Example 1

The methodology outlined in this entry has been used in a number of studies to measure levels of conceptual understanding in science. In one study (Bowden et al., 1992; Ramsden et al., 1992) 14 physics problems were developed, each designed to explore students's conceptual understandings of an aspect of mechanics. Students attempted each problem during an interview and their explanations were tape recorded and transcribed. Interview transcripts were then analyzed and students' responses were grouped to form a hierarchy of qualitatively different levels of understanding of the phenomenon addressed by each question.

One of these 14 questions addressed students' understanding of speed as a relative quantity:

> Martha and Arthur are running along a straight road at constant speed. Arthur is ahead of Martha. Arthur's speed is less than Martha's speed. How far must Martha run before she catches up to Arthur, and how long will this take her?

A detailed analysis of several dozen student transcripts led to the identification of six different strategies these students used to answer this question. These have been combined in Figure 1 to define four ordered outcome categories for this question. The most sophisticated understanding is reflected in the approach described in category 3. Students assigned to category 0 show very little understanding and may refer to the "acceleration" of Martha or rely on equations that relate the relevant variables but that are inappropriate. Excerpts from students' verbal and written explanations illustrating these four categories are shown in Figure 2.

An important difference between the four categories in Figure 1 and illustrated in Figure 2 and more usual approaches to defining ordered levels of response to an item is that, rather than being defined in terms of stages in the solution of a problem, these categories reflect qualitatively different strategies and understandings. In this study, ordered categories were defined for each of the 14 physics questions and used to construct measures of student understanding in this area of physics.

3

Students who take this approach solve the problem by focusing on relative speed and relative distance. They focus on the "gap" as the distance that Martha must run. She closes this gap at the "catching speed". The ground appears to be automatically conditioned out of consideration, indicating a sophisticated understanding of speed as a relative quantity. Most students in this category solve the problem using:

$$\text{time to close gap} = \frac{\text{size of gap}}{\text{catching speed}}$$

2

Students who take this approach consider the motion of each runner separately with respect to the ground. Typically, they set up two separate equations of motion and attempt to solve those two equations simultaneously. Alternatively, they may solve the problem graphically by finding the intersection of two separate straight lines on a distance–time graph. Almost all students taking this approach understand that Martha and Arthur run for the same time t and use this fact in the solution. Some students are able to arrive at a solution: e.g.,

$$\text{distance Martha runs} = \textit{distance Arthur runs} + \textit{gap}$$

i.e., $V_m \times t = V_a \times t$ $+ \textit{gap}$

Others fail to solve the problem because they do not incorporate the initial distance between Martha and Arthur: e.g., time Martha runs = time Arthur runs

i.e., $t = d_m / V_m = d_a / V_a$.

1

Students who take this approach focus on the motions of Martha and Arthur separately and, rather than attempting to derive a general algebraic or graphical solution, adopt a "trial-and-error" approach. Typically, they divide the continuous motion into discrete pieces and consider the relative locations of the two runners after fixed intervals of time or after one of them has run a particular distance. This may or may not lead to a solution of the problem.

0

A small number of students when presented with this problem in written form fail to display an understanding of the problem or of an appropriate approach. These students may produce one or more equations of motion and attempt to substitute into these equations, but they show little evidence of understanding and invariably become lost. This category includes all students who fail to demonstrate the levels of understanding reflected in categories 1, 2 and 3.

Figure 1   Response categories for physics question.

"The difference in velocities (in m/sec) is how much Martha would gain on Arthur if the following equation is used:

V = Martha's velocity − Arthur's velocity.

3 Once we know how much Martha would gain on Arthur per second, we can then divide the original distance between them by V to find out how long it would take Martha to catch Arthur, in seconds:

$$\text{time taken to catch Arthur} = \frac{\text{original distance between runners}}{\text{V}}$$

2 "In the time t, Martha will tavel $V_m . t$ and Arthur will travel $V_a . t$ . . . But given that Martha has to travel an extra d, we'll take that away from the distance and then they'll be in the same place:

$$V_m . t - d = V_a . t\text{''}.$$

1 "Work out the time taken for Martha to travel the initial distance (S_i) between them. i.e., time = S_i / V_m. Use this time to find out how far Arthur ran. Then find out how long it takes Martha to run this extra distance. Then find out how far Arthur ran in this time.

Keep doing this until they have run the same distance (i.e., Martha has caught up to Arthur)".

0 "v = speed Martha, u = speed of Arthur,
s = distance between the two, t = time.

$$s = ((v + u)/2) . t\text{''}$$

Figure 2 Student responses illustrating the categories.

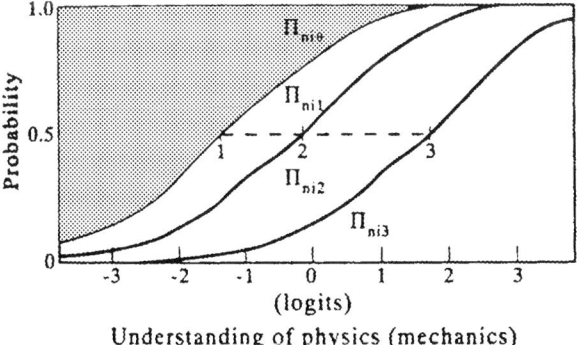

1.0
0.5
0
Probability
Π ni0
Π ni1
Π ni2
Π ni3
1
2
3
-3
-2
-1
0
1
2
3
(logits)
Understanding of physics (mechanics)

Figure 3 Modeled probabilities for "Martha and Arthur" question.

Figure 4   Response categories and levels for a "living things" question.

	Level
Responses that include a list containing a number of the usually identified characteristics of living things. The answer need not be complete in biological terms	3
Responses that focus on movement	2
Responses that focus on growth	2
Responses that focus on the appearance of the object	1
Responses that suggest *poking at* or touching the object with a stick. A reason for poking or touching the objects is not provided	1
Uninterpretable responses	0

The application of the partial credit model (1) to this physics question is illustrated in Figure 3. This figure shows the modeled relationship between the variable being measured by the set of 14 questions and the response probabilities for the "Martha and Arthur" question. It can be seen that, under the model, the greater a student's general understanding in this area of physics, the more likely the student is to display a sophisticated understanding of speed as a relative quantity.

The points labeled 1, 2 and 3 in Figure 3 mark boundaries between the four categories defined for this question. Measures of conceptual understanding on the horizontal variable in Figure 3 can be referred to the category boundaries for all 14 items to provide an interpretation of the kinds of understandings that typify each measure. A student with an estimate of 1.0 logits on this scale will probably respond in category 1 but below category 3 (i.e., in category 2) to the "Arthur and Martha" question, for example.

6.2   Example 2

A second application of the methodology described in this entry is found in the Victorian Science Achievement Study by Adams *et al.* (1990) of students' understandings of a number of different science phenomena (see also Doig and Adams, 1991). The Victorian

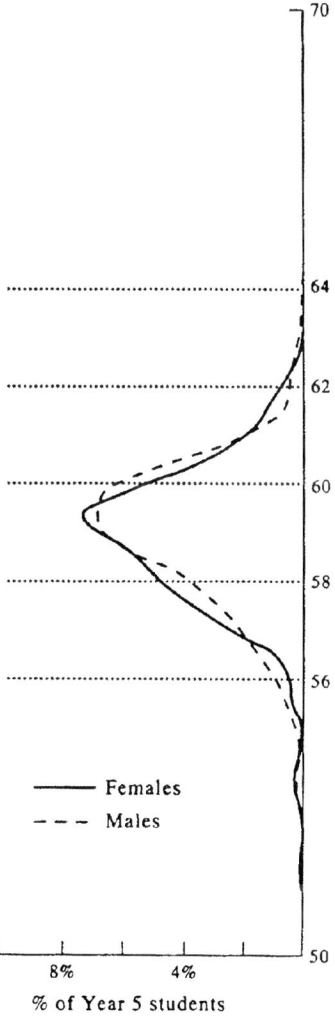

Figure 5   The "living things" variable.

Students are able to use scientific terminology in supporting evidence; for example, to make distinctions between living and non-living things. (64)

Some evidence that students are able to relate observations, provide multiple reasons to justify their responses, and draw inferences from evidence provided. (62)

Students have some basic scientific knowledge such as humans are animals. (60)

Students can provide justifications for their responses but that justification is often unscientific and direct.

Student understandings are unscientific. They provide responses based upon folklore. Little or no justification or supporting evidence is provided for responses. (58)

Students provide responses that are uninterpretable. (56)

Science study began with a detailed review of the science education research literature to establish the kinds of student misconceptions reported in earlier studies. This review of previous research provided a guide to the construction of ordered response categories for each of the questions used in the study.

The questions used in the Victorian Science study were based on short comic-strip sequences around particular themes (e.g., living things, earth and space, force and motion). Each question invited students to respond to the discussion in the comic strip and to explain some aspect of science in their own words. An example of one of the open-ended questions used with fifth-grade students is shown at the top of Figure 4.

The range of responses that students might give to this question can be anticipated to some extent from previous research evidence (e.g., Bell, 1981; Trowbridge and Mintzes, 1988). The six different categories of response in Figure 4 were identified in the written responses of fifth-grade students in the Victorian study. On the right of Figure 4, these six responses have been clustered to form four "levels" of performance on the question. The lowest level of response is labeled 0, the highest is labeled 3. In this way, all responses to the question were assigned to one of four ordered response levels.

Students' responses to all eight Living Things questions were then analysed using the partial credit model (1) and the results of the analysis were used to construct the variable shown in Figure 5. The text below Figure 5 describes increasing levels of understanding of the characteristics of living things.

On the left of Figure 5 are the distributions of male and female fifth-grade students' levels of performance. It can be seen that very few fifth-grade students demonstrated an ability on this set of questions to "relate their observations", to "provide multiple reasons to justify their responses", or to "draw inferences from evidence".

The methodology described in this entry and illustrated in these two examples can be used to mark out a measurement variable in terms of qualitatively different levels of understanding. Measures derived and interpreted in this way are capable of providing a more instructionally-useful basis for charting student learning than some more traditional testing methods.

References

Adams, R. J., Doig, B. A. and Rosier, M. 1990. *Science Learning in Victorian Schools*. Australian Council for Educational Research, Hawthorn, Victoria.
Bell, B. F. 1981. When is an animal not an animal? *J. Biol. Ed.* **15**(3), 213–18.
Bloom, B. S., Hastings, J. T. and Madaus, G. F. 1971. *Handbook on Formative and Summative Evaluation of Student Learning*. McGraw Hill, New York.
Bowden, J. *et al.* 1992. Displacement, velocity, and frames of reference: Phenomenographic studies of students' understanding and some implications for teaching and assessment. *Am. J. Physics* **60**, 3.
Caramazza, A., McCloskey, M. and Green, B. 1981. Naive beliefs in "sophisticated" subjects: Misconceptions about trajectories of objects. *Cog.* **9**(2), 117–23.
Champagne, A. B., Klopfer, L. E. and Anderson, J. H. 1980. Facotrs influencing the learning of classical mechanics. *Am. J. Physics* **48**(12), 1074–9.

Chi, M. T. H., Feltovich, P. J. and Glaser, R. 1981. Categorization and representation of physics problems by experts and novices. *Cognit. Sci.* **5**, 121–52.

Clement, J. 1982. Students' preconceptions of introductory mechanics. *Am. J. Physics* **50**(1), 66–71.

Dahlgren, L.-O. 1984. Outcomes of learning. In: Marton, F., Hounsell, D. and Entwistle, N. (eds.) 1984. *The Experience of Learning*. Scottish Academic Press, Edinburgh.

diSessa, A. 1982. Unlearning Aristotelian physics: A study of knowledge-based learning. *Cognit. Sci.* **6**, 37–75.

Doig, B. A. and Adams, R. J. 1991. The Victorian Science Achievement Study: Implications for practice. *Aust. Sci. Teach. J.* **37**(4), 25–31.

Driver, R. and Easley, J. 1978. Pupils and paradigms: A review of literature related to concept development in adolescent science students. *Stud. Sci. Educ.* **5**, 61–84.

Gunstone, R. and White, R. 1981. Understanding of gravity. *Sci. Educ.* **65**(3), 291–9.

Johansson, B., Marton, F. and Svensson, L. 1985. An approach to describing learning as change between qualitatively different conceptions. In: West, L. H. and Pines, L. A. (eds.) 1985. *Cognitive Structure and Conceptual Change*. Academic Press, Orlando, Florida.

Larkin, J. H. 1983. The role of problem representation in physics. In: Gentner, D. and Stevens, A. (eds.) 1983. *Mental Models*. Erlbaum, Hillsdale, New Jersey.

Lybeck, L. *et al.* 1988. The phenomenography of "the mole concept in chemistry". In: Ramsden, P. (ed.) 1988. *Improving Learning – New Perspectives*. Kegan Paul, London.

Marton, F. 1981. Phenomenography – describing conceptions of the world around us. *Instr. Sci.* **10**(2), 177–200.

Masters, G. N. 1982. A Rasch model for partial credit scoring. *Psychometri* **47**(2), 149–74.

McCloskey, M. 1983. Naive theories of motion. In Gentner, D. and Stevens, A. (eds.) 1983. *Mental Models*. Erlbaum, Hillsdale, New Jersey.

McCloskey, M., Caramazza, A. and Green, B. 1980. Curvilinear motion in the absence of external forces: Naive beliefs about motions of objects. *Science* **210**(4474), 1139–41.

McDermott, L. C. 1984. Research on conceptual understanding in mechanics. *Physics Today* 1–10.

Nesher, P. 1986. Learning mathematics: A cognitive perspective. *Am. Psychol.* **41**(10), 1114–22.

Osborne, R. J. and Wittrock, M. C. 1983. Learning science: A generative process. *Sci. Educ.* **67**(4), 489–508.

Posner, G. J., Strike, K. A., Hewson, P. W. and Gertzog, W. A. 1982. Accommodation of a scientific conception: Toward a theory of conceptual change. *Sci. Educ.* **66**(2), 211–27.

Ramsden, P. *et al.* 1992. Phenomenographic research and the measurement of understanding: An investigation of students' conceptions of speed, distance and time. *Int. J. Educ. Res.* **19**(3), 301–16.

Säljö, R. 1984. Learning from reading. In: Marton, F., Hounsell, D. and Entwistle, N. (eds.) 1984. *The Experience of Learning*. Scottish Academic Press, Edinburgh.

Trowbridge, D. E. and McDermott, L. C. 1981. Investigation of student understanding of the concept of acceleration in one dimension. *Am. J. Physics* **49**(3), 242–53.

Trowbridge, D. E. and Mintzes, J. J. 1988. Alternative conceptions in animal classification: A cross age study. *J. Res. Sci. Teach.* **25**(7), 547–71.

Viennot, L. 1979. Spontaneous reasoning in elementary dynamics. *Eur. J. Sci. Educ.* **1**(2), 205–21.

White, B. Y. 1983. Sources of difficulty in understanding Newtonian dynamics. *Cognit. Sci.* **7**(1), 41–65.

White, B. Y. and Horwitz, P. 1987. Thinker Tools: Enabling children to understand physical laws. Proc. 9th Annual Conf. of the Cognitive Science Society, Erlbaum, Hillsdale, New Jersey.

Wright, B. D. and Masters, G. N. 1982. *Rating Scale Analysis*. MESA Press, Chicago, Illinois.

22   Issues in Educational Measurement

J. P. Keeves and G. N. Masters

The issues arising from the developments that have occurred in item response theory over the past 30 years lie not in the further development of theory, but in the application of the abundant theoretical ideas and relationships that have already been advanced (Fischer and Molenaar, 1995; van der Linden and Hambleton, 1997). The power of these principles of measurement is limited by the availability of computer programs, with clearly stated guidance for their use, although this shortcoming is rapidly being rectified. There is also a lack of widespread experience in their application in research and practice in the fields of education, psychology and the social sciences.

1   Missing responses

The operations of measurement in education where large numbers of persons are involved, is plagued by the existence of missing data. Both items and persons and the interrelations between items and persons must be considered in the proper handling of the missing data. If the missing data were to occur through completely random processes, then the procedures of item response theory in which the performance of a particular person relative to a particular item is treated as the unit of analysis, permits estimation with incomplete data. However, except for printing and processing errors, the presence of missing data does not occur at random. It is not only a characteristic of particular persons but also of particular items as well as the interrelations between persons and items. As a consequence, any results based on data in which there are missing values are biased to the unknown extent that the persons and items with missing values occur within the data set.

The data collected from the administration of a test or attitude scale contains missing values which arise in two distinct ways. First, there are missing data that occur in the body of responses, with neighbouring responses being present. Such missing data commonly arise because the respondents do not know the answer to a particular question, or are unclear of their emotional reaction to a particular attitude statement. These missing data are referred to as "omitted responses". Secondly, there are missing data that occur at the

268

end of a set of responses, simply because the respondents were not allowed sufficient time to complete the tasks required of them. These missing data are referred to as "not reached responses". The presence of not-reached responses implies that items in tests and statements in attitude scales are responded to in a serial order.

The issue of missing responses in tests has been fully addressed by Longford (1994). However, the issue is of particular significance in testing programs in which the test-takers have little or nothing at stake, and are inclined to refrain from answering where considerable effort is involved. Under such circumstances, omission and not reaching are quite frequent because the test-takers have little or no motivation to respond.

> No amount of effort or ingenuity expended in instructing and motivating test-takers can ensure that they diligently attend to each item and respond accurately to background and experiential items, and respond to cognitive items to the best of their ability. This problem is particularly prevalent when the test-takers have little or no stake in the outcome of the test administration. (Longford, 1994, p. 2)

Much has been written (for example, Ludlow, 1994) on the problems of missing data where omitted and not-reached responses occur, but without clear resolution of the issues. Furthermore, where in testing programs it has been assumed that such missing data have occurred at random, or where they have been ignored, the results of the testing program have sometimes been strongly challenged and the results condemned. There is little doubt as Longford (1994, p. 12) has pointed out that "not reaching is more frequent among those who omit more", and that "less able test-takers are more likely to omit". It is, however, necessary to consider these problems of missing data with respect to two distinct aspects of analysis using item response theory, namely, scale calibration and scoring.

A further aspect of the missing data problem arises when alternative forms of a test are administered to different subgroups of persons. This procedure is widely used in testing programs to extend the curriculum coverage of the tests without increasing the time required for responding. Under these circumstances it is essential that the alternative forms of the test are allocated randomly to the persons under survey. Failure to distribute the alternative forms at random both between schools and within schools is likely to introduce bias and should be avoided.

Since it must be accepted that missing data, whether as omitted responses or not-reached responses, does not occur at random, it is inappropriate to ignore the missing data in calibration even though this procedure is widely employed. Moreover, while there is the propensity for less-able persons to fail to respond to test items, there is also the tendency for more cautious persons, and not necessarily the less able, to be reluctant to guess as well as to work more slowly and not complete a test. Consequently, the consideration of missing data as a wrong response in a test would seem to be highly unsatisfactory in the operation of scale calibration. In the calibration of responses to an attitude scale, provided the extent of missing data by a person in responses to an attitudes statement does not exceed, perhaps a value of 20 per cent, it would seem appropriate to consider the missing data as falling in the uncertain category. However, this procedure may cause problems associated with reversed thresholds in calibration.

If sufficient numbers of persons are available, the only practicable approach in the processing of achievement tests is to use, in the calibration stage of processing, only those persons who have responded to all items or statements. Since the calibrated difficulty levels of the items are independent of the ability or performance levels of the persons responding, this restriction on the sample of the persons used in calibration is of no consequence. Other procedures such as processing the omitted and not reached items as wrong does influence the average difficulty level of the items, because different items are influenced to different extents by this procedure.

In the operation of scoring student responses, the ignoring of omitted and not-reached responses is fully possible in the estimation of scale scores and in some ways may be preferred to considering such responses as wrong. The ignoring of missing data in scoring commonly leads to higher mean scores and smaller standard deviations on a calibrated scale, while the treatment of missing data as wrong leads to lower mean scores and larger standard deviations. The latter procedure would seem more easily explained to parents, teachers and students in a testing program, since it is hard to argue that a student who omitted items in a test or did not complete a test could obtain a higher score than a student who answered all items. Lietz (1996) and Hungi (1997) have investigated these issues and sought to examine the strength of alternative procedures. Moreover, it is also evident that the larger standard deviation of scores obtained from the scoring of missing data as wrong leads in some situations to marginally larger correlations with key predictor variables, implying greater strength and validity in such results.

The practical issues involved in the treatment of missing data in both calibration and scoring have not been sufficiently explored for full resolution of the problems to have been achieved. This issue would seem of considerable practical significance for further work to be done until consensus is reached on the most appropriate procedures to be employed.

2   Dimensionality

The practical issue of dimensionality arises, particularly in North America, where Mathematics and Science are taught in a segmented way at the middle and upper secondary school level as Arithmetic, Algebra, Geometry, Trigonometry and Calculus in Mathematics, and as Biology, Chemistry, Earth Science and Physics in Science. Likewise, in a field like Reading Comprehension, performance is viewed in terms of identifiably separate skills, or in terms of the domains of narrative prose, expository prose, and documents. The issues in these and many more cases is whether Mathematics, Science and Reading Comprehension could be considered to be unidimensional fields, and the calibration of a single scale could be meaningfully undertaken, or whether several domains exist within each field requiring a multidimensional approach and the rejection of the construction of a single scale in each field. There is the further question of whether unidimensionality is a matter of degree, with no clear resolution of the issue possible. Alternatively, it is possible that both a single factor as well as orthogonal multiple factors can be considered to exist in data associated with a particular field.

Bejar (1983, p. 30) has pointed out that unidimensionality is an artefact, constructed in investigation "to account for the covariation among responses to a set of items". Furthermore, Bejar has argued that:

> unidimensionality does not imply that performance on the items is due to a single psychological process. In fact, a variety of psychological processes are involved in the act of responding to a set of items. However, as long as they function in unison – that is, the performance on each item is affected by the same processes and in the same form – unidimensionality will hold. (Bejar, 1983, p. 31)

However, the presence of a degree of the essential unidimensionality in a set of items indicates that the persons responding to those items are influenced by common psychological processes as a consequence of common learning experiences. Moreover, in the absence of a degree of unidimensionality it would not seem meaningful to calibrate a scale or to calculate scores for person or items on that scale. In order to use the Rasch model it is a requirement that a high degree of the essential unidimensionality should hold among the persons and the items to which they respond.

Confirmatory factor analysis is now readily carried out using the LISREL8W computer program (Jöreskog and Sörbom, 1993) in order to examine alternative factor structures among sets of items and the resulting evidence can be used to infer the underlying dimensionality of the set of items (Marsh and Hocevar, 1985; Hattie, 1985). More recently, Bentler and Houck (1996) have pointed out that confirmatory factor analysis is equivalent to performing a two-parameter item response theory analysis. Consequently, the results of a confirmatory factor analysis can also be employed to compare alternative models and to indicate how the data would behave if the more rigorous one-parameter model (or Rasch model) were fitted to data. Generally, the root mean square error of approximation (RMSEA) (Gustafsson and Stahl, 1996, p. 88) is employed as an index in a test of close fit, which is not dependent on sample size.

Alternative models involving a single factor, several correlated factors and a nested structure in which a single factor is set orthogonal to several correlated factors can be compared for goodness of fit. It is commonly found that a several correlated factors model provides a better fit than the single factor model, and that the nested model provides a better fit than the several correlated factors model. The nested model indicates that both the single factor and the orthogonal several correlated factors can be employed to describe the relationship within the set of items. An examination of the variance associated with the single factor compared to the total variance associated with the several correlated factors provides evidence of the relative strength of the unidimensional model when compared with a multidimensional model (Mohandas, 1996; Hungi, 1997). If the variance extracted by the single factor exceeds the variance extracted by the several correlated factors which are orthogonal to the single factor, it would seem clear that unidimensionality is to be preferred for the data set. If the reverse were true with the variance of the several correlated factors exceeding the variance extracted by the orthogonal single correlated factor, then the existence of unidimensionality must be considered to be questionable, even though the set of items might well appear to satisfy the requirements for fitting the Rasch model. In the use of the Rasch model, greater consideration should be given to the examination of the dimensionality of the set of items, prior to testing whether the items provide evidence of fit to the Rasch model. Such analyses might also assist in determining whether the two-parameter or the three-parameter model should be used in preference to the one-parameter Rasch model in particular circumstances.

Nevertheless, it must be recognized that the use of confirmatory factor analysis is only a first step in the examination of a data set for unidimensionality or multidimensionality. It is subsequently necessary, to examine both persons and items for fit to the Rasch model. Unfortunately there are no clearly acknowledged criteria for acceptance of fit to the Rasch model, and the use of the Rasch model to obtain strong measures is still frequently challenged. As a consequence, it is essential to provide all possible evidence to support the use of the Rasch model when it is employed for measurement in educational and psychological research. The efforts of research workers need to focus on these problems until clarification and agreement are reached.

3 Methods of equating

In the use of the one-parameter or Rasch model, three methods of equating have emerged as appropriate procedures for combining together the two sets of items that form different tests, whether the equating is being undertaken horizontally with the person performance and item difficulty at approximately the same scale levels, or vertically with person performance and item difficulty of the tests at very different scale levels. The first procedure is the common item threshold difference method, in which the two sets of items, that have a sufficient number of common items, which have been purposefully chosen, are separately calibrated, and the average threshold difference between the two subsets of common items is calculated as the equating constant between the two scales. Since the items are fixed as common items, not only can the mean difference be calculated but the error associated with that mean difference can also be estimated.

The second method involves the calibration of one set of items to form a scale. The threshold values of those items in the calibrated set that are common to a second set are employed in the calibration of the second set. These items are anchored at those threshold levels, with the items in the second set thus being scaled at levels corresponding to those items in the original or anchoring set. This procedure is referred to as the anchor item equating method.

The third procedure is a concurrent equating method. With the construction of computers with large storage capacity, it has become possible to combine together two or more data sets with common items or with common persons in such a way that a single calibration of the several sets of items and the corresponding sets of person data can be undertaken. In this method the persons and items in the several data sets are brought to a common scale. While the errors of equating in this way have not been estimated, there is evidence that this procedure provides more consistent and stronger measures of the two sets of items and persons that have been equated (Baker and Al-Karni, 1993; Mohandas, 1996).

The file merging procedures now available permit the concurrent equating procedure to be readily applied. Moreover, common person equating as well as common item equating can be used. Nevertheless, the errors involved in equating by this method remain unestimated and since the magnitude of equating errors is of importance this problem awaits further work.

4 Significance

The use of statistical significance is undergoing systematic reexamination in educational and psychological research. Greater consideration is being given to the size of effect and the pattern of results. In fields of inquiry where carefully controlled experiments can be readily conducted and replicated it is possible to estimate appropriate sample sizes for the detection of an expected effect through power calculations, but, the question of "how many persons?" cannot be so readily answered in educational research, where even interventions are difficult to undertake and where survey research and investigations must largely prevail. Small effects with a high level of statistical significance can be easily obtained if large samples are used. However, the practical significance of a small effect is often open to question. Moreover, effects from survey research, and investigations are sometimes difficult to replicate. As a result there is a growing interest in the assessment of the size of an effect and the pattern of the results.

In the commonly employed tests for detection of item fit and person fit that are used in item response theory analyses to determine if a particular item or a particular person fits the Rasch model, the examination of fit remains largely dependent on sample size and tests of statistical significance. The alternative procedure would seem to be to employ a test of deviation from fit that was independent of sample size, being related to the magnitude of the deviation. While the use of the weighted or infit mean square statistic for item fit seeks to shift from a test of statistical significance to a range test, the range commonly employed is only justifiable by experience and not by argument. Further problems arise because items are commonly clustered in subsets and the items are not independent of one another.

Furthermore, in most samples of persons that are employed in educational research, the simple random sample tests of statistical significance are largely inappropriate because a cluster sample of students within classes, or students within schools is used to obtain large numbers of students from relatively few schools or classes. Farish (1984) is the only known publication that examines the consequences of cluster sample design for the fit statistic and parameters estimated in item response theory analysis. While students respond to tests and attitude questionnaires, these students are clustered in schools and classes and are selected for study as intact schools and classes, with the results that the students do not represent independent sampling or treatment units. Under these circumstances the use of simple random sample based tests is commonly greatly in error. Computer programs that employ jackknife and balanced repeated replication have recently become available (Brick *et al.*, 1996) for the calculation of mean values, proportions and regression coefficients. The use of such procedures needs to become more widely employed if significance tests continue to be widely used in measurement with item response theory.

5   Efficiency of estimation

With the development of a range of computer programs that can be employed for the estimation of the person and item parameters in Rasch scaling, it is acknowledged that different computer programs serve different purposes. In large scale testing programs, particularly those from which student performance estimates are used for reporting to individual students, teachers and parents it is necessary to employ highly efficient estimation procedures. Such procedures must report performance estimates for persons with small standard errors. Where group mean performance is of interest the sampling errors are commonly greatly in excess of the estimation or measurement errors, particularly since it is customary to sample schools or classes within schools at the first stage of sampling and then to use intact school or classroom groups at the second stage. However, in reporting an estimate of student performance in which the standard error is equivalent to one year of student learning, such errors would seem unacceptably large. Under these circumstances, greater consideration must be given to the use of more efficient estimation procedures, and more items being employed in the estimation of student performance. The possibility of using computer adaptive testing procedures in order to reduce the standard errors of estimation without increasing the length of a test, and identifying the maximum acceptance size of errors for the reporting of student performance needs to be investigated.

6   The reliability index

Under classical test theory the use of the index of the reliability of a test and of the sets of scores derived from a test has been prominent. The index of reliability for a set of scores has also been used to provide an estimate of the errors of measurement associated with those scores. Moreover, the reliability index has been used more generally to set an upper bound for the validity of the test. However, the value of this index is dependent on the spread of ability or performance of the persons responding to the test, as well as on the properties of the items included in the test. As a consequence the reliability index only has meaning in the particular situation in which the test was administered and only applies to the scores derived from that administration. As a general index to describe the properties of a test or instrument, the reliability index is of doubtful value.

In item response theory this index of reliability is largely redundant since as Andrich (1982) has pointed out the emphasis is on the fit of the items to the Rasch model, and on the standard errors of the estimates of the difficulty levels of individual items and the performance levels of individual persons. It is, nevertheless, sometimes useful to know the proportion of the total variance associated with the group of persons responding to a test that can be regarded as error variance and the proportion that can be considered to be parameter variance. In regression analysis and other multivariate analysis procedures it is the parameter variance that is explained in analysis, and the amount of parameter variance available for explanation provides some indication of the worthwhileness of the analysis. The index providing this information is referred to as the person separation index within item response theory.

7   The person separation index

The proportion of the total variance of the scale estimates for persons (β_n) that is associated with parameter variance is given by the person separation index (S).

From an analyses of variance of a set of scores the person separation index can be estimated

$$S = \frac{\text{Person mean square} - \text{Mean sum of squares of errors of measurement}}{\text{Person mean square}}$$

where

$$\text{Person mean square} = \sum_{n=1}^{N} \frac{(\beta_n - \beta)^2}{N - 1}$$

However, the errors of measurement of persons can under item response theory be estimated separately for each person, and can be squared and summed to calculate a mean sum of squares of errors of measurement that could differ from the mean square error term obtained from an analysis of variance.

The term for the mean squares sum of errors of measurement is given by

$$\sum_{n=1}^{N} e_n^2$$

where

β_n = scale estimate for person n
β = mean value of the scale estimate
e_n = estimated error of measurement for person n
N = total number of persons.

The values of the person separation index are similar to the coefficients of reliability but the error of measurement is conceived somewhat differently in the calculation of the person separation index when compared with the calculation of error variance in classical test theory.

A similar coefficient can be calculated for items as the Item Separation Index (I).

Where the values of the person separation index and the item separation index are large, the persons and items are spread across the number scale and the estimation of the scale parameters is undertaken with greater efficiency. This is evident from the fact that where the separation coefficients are high the errors of the measurement are small compared to the variability between persons or items, and this implies better estimation of the person and item parameters. An increase in the numbers of persons and items also increases the values of the separation indexes, reduces the sizes of the errors of measurement, and thus increases the efficiency of estimation.

8   The fit of items and persons to the Rasch scale

The errors of measurement and hence the separation index are commonly calculated from formulas under the requirement that the items and the persons conform to the Rasch scale. If, however, the items and the persons do not conform to the Rasch scale, not only are the parameter estimates distorted, but the estimates of error do not provide evidence of the distortion of measurement that has occurred. Under these circumstances the rules and procedures established for the fit of items and persons to the Rasch scale become of very considerable significance.

Improvement in measurement can only occur if more stringent conditions are laid down for the fit of items and persons to the requirements specified by the logistic function. At the present time decisions regarding item and person fit are made largely by inspection of:

(a) the fit of fractile groups to the item and person characteristic curves;
(b) the magnitudes of fit indexes within a somewhat arbitrary range of values expressed as either a t-statistic or a χ^2 statistic;
(c) the size of the probability levels associated with the t-statistic and the χ^2 statistic, which might have failed to allow for the clustering of items within sub-testlet, or the clustering of persons within groups that are sampled at a first stage of sampling;
(d) for items, the item discrimination indexes, obtained as the point-biserial correlations between the response of a person to an item and the person total raw score;

(e) for items, the infit or weighted mean square statistic, that is independent of sample size, within a specified range; and

(f) for persons, the outfit or unweighted mean square statistic, that is also independent of sample size, within a specified range.

It should be noted, however, that some of these statistics and their associated probability levels are highly dependent on sample sizes either for persons or for items. Moreover, the different estimation procedures that are employed in the different computer programs indicate different items and persons as not fitting the Rasch model, particularly where there is generally poor fit of items and persons to the model.

Under these circumstances there would appear to be a general lack of agreement as to what constitutes fit to the Rasch model, particularly for items that discriminate strongly between persons. Items that fail to discriminate satisfactorily between persons are generally easier to identify as unsatisfactory items.

9   Problems found in the measurement of attitude

Problems are encountered in the measurement of attitudes that arise from two sources. First, statements of attitude that require a bipolar response, if they lie near the middle of the scale range rather than at the extremes, can be disagreed with if positive statements or agreed with if negative statements, on two grounds. The person responding may be located either well above or well below the statement on the attitude continuum. Analysis of data that involves such statements with Rasch scaling programs using the one parameter logistic function, tends to show acceptable fit but inconsistent threshold values. This problem can be overcome by using an unfolding procedure (Andrich, 1995) and the RUMMFOLD computer program (Luo, Andrich and Sheridan, 1997) which employs the hyperbolic cosine function instead of the logistic function. Secondly, responses in the undecided or uncertain category, can exhibit similar effects of inconsistency in threshold values, which can only be resolved by the listwise exclusion of those persons who have responded in this way to an item in the calibration of the attitude scale. However, those persons can be included for scoring, once calibration has been completed.

10   Interpreting partial credit parameters

A similar problem involving inconsistent threshold values arises in the use of the partial credit model. In order to achieve statistical separation of person and item parameters and hence specific objectivity, the Rasch partial credit model (see *Partial Credit Model*) is written as the log odds of person n and ability β_n responding in category x rather than in category $x - 1$ of item i. In other words, the basic model is formulated in terms of each *pair* of adjacent response categories: $\beta_n - \delta_{ix} = \log(P_{nix}/P_{nix} - 1)$. If δ_{ix} is reparameterised as $\delta_i + \tau_x$, then the partial credit model becomes the rating scale model.

The formulation of the model in terms of each pair of adjacent categories can be contrasted with *cumulative* formulations (e.g., Samejina, 1997) which consider all response categories $(0, 1, \ldots, x - 1)$ up to some 'threshold' γ_{ix} and all categories $(x, x + 1, \ldots, m)$ beyond that threshold:

$$\beta_n - \gamma_{ix} = \log \left(\sum_{j=x}^{m} P_{nij} \bigg/ \sum_{j=0}^{x-1} P_{nij} \right)$$

Because they are locally rather than cumulatively defined, there is no algebraic constraint on the order of the parameters $\delta_{i1}, \delta_{i2}, \ldots, \delta_{im}$ in the partial credit model or the parameters $\tau_1, \tau_2, \ldots, \tau_m$ in the rating scale model. (This is in contrast to cumulatively defined parameters which, by definiton, are constrained to be ordered $\gamma_{i1} < \gamma_{i2}, \ldots, < \gamma_{im}$). The absence of such a constraint means that when the partial credit and rating scale models are applied, estimates for the parameters (δ_{ij}) and (τ_j) can be in any order. This raises the question of how test developers and data analysts should respond to the *order* of these parameter estimates.

One view is that the order of the parameter estimates in the partial credit and rating scale models is not an indication of the extent to which data conform to the measurement model or of the extent to which objective measurement is possible. When estimates occur in an order other than $\delta_{i1} < \delta_{i2}, \ldots, < \delta_{im}$, this does not indicate that the response categories are not functioning as ordered levels of performance and response as intended, but usually indicates that some responses are relatively uncommon. In certain cases, this may indicate the need to revise the scoring scheme or rating categories to ensure a more even distribution of category use. When estimates occur in the order $\delta_{i1} < \delta_{i2}, \ldots, < \delta_{im}$, they mark out a region of 'most probable response' for each response category, but this is not the only way to map ordered response categories on to a measurement variable. An alternative approach – consistent with the approach used by Thurstone to mark out measurement variables, and available through the BIGSTEPS, QUEST and CONQUEST programs – is to identify the point on the variable at which

$$\sum_{j=0}^{x-1} P_{nij} = \sum_{j=x}^{m} P_{nij}$$

and to use the resulting cumulatively defined 'thresholds' to mark out and interpret the variable.

An alternative view is that, if parameter estimates are not ordered $\delta_{i1} < \delta_{i2}, \ldots, < \delta_{im}$, then the model itself is violated, and objective measurement may not be possible. Under this view, the item parameters $\delta_{i1}, \delta_{i2}, \ldots, \delta_{im}$ are conceptualised as 'thresholds' which must be ordered on the measurement variable. This view also contends that orders other than $\delta_{i1} < \delta_{i2}, \ldots, < \delta_{im}$ can be the result of differences in discrimination, and that strong and consistent measurement requires equal discrimination.

These two views warrant both theoretical and empirical investigation in order to determine whether differences between the views can be resolved, since widespread use is now being made of the partial credit model. Moreover, these two views lead to different approaches to locating item thresholds on a scale of performance.

11   Misfitting persons

Because it is common to employ fewer items than persons in calibration, any deviations from the Rasch model are of greater consequence for items than for persons. Nevertheless, it would seem to be assumed that since there are many more persons than items involved in the calibration of most Rasch scales, the distortion of measurement arising from misfitting items is potentially more serious than the distortion arising from misfitting persons. However, as greater use is made of Rasch scaling in educational and psychological research studies, more attention must be given to the consequences of employing misfitting persons as well as items in the calibration of a scale. This should lead to the discarding of misfitting persons as well as those who have obtained zero and perfect scores in the Rasch scaling.

Since it is the relationship between persons and items, given by $(\beta-\delta)$, and the ability of a person relative to the difficulty of an item that is employed in estimation, when the absolute value of this difference becomes large the strength of estimation declines. As a consequence estimation could be improved without rejecting persons or items from the estimation process using maximum likelihood procedures by weighting down these paired comparisons where the magnitude of the expression $(\beta-\delta)$ is large, and weighting up these paired comparisons where the expression $(\beta-\delta)$ is small. An alternative procedure would be to excude from consideration those paired comparisons where $|(\beta-\delta)|$ exceeds a value, possibly 2. However, if these item person pairs did not occur at random within the data set, bias might well be introduced into the estimation process, by the exclusion of certain paired comparisons in this way.

12   Conclusion

The improvement of measurement in educational and psychological research is not an end in itself. It is a means to the end of detection of stronger relationships that are of theoretical and practical interest. Both fields of research are concerned with the stability and change in human characteristics that are central to the processes of teaching and learning. Inquiry into the factors influencing learning has been constrained by the difficulties encountered in measuring change in performance and attitudes over time. In order to avoid the problems of unreliability of difference scores, it is now acknowledged that performance and learning outcomes should at least be measured at three points in time. However, the effects of practice on performance distort such measurements if the same instrument is used to obtain successive measurements. The construction of a scale of performance that is independent of the items and persons employed in the calibration of the scale, permit different instruments to be used on different occasions, provided these different instruments can be equated to a common scale. The ease with which this can be achieved through Rasch scaling, and the construction of an interval scale by means of which change can be measured in a meaningful way at different levels on the scale opens up the field of research involving investigation into the problems of learning. Furthermore, the construction of scales of performance has the capacity to change the way in which teaching takes place and the meaning of the results of assessment that are reported back to students, parents and teachers from a testing program. No longer are students compared on a scale of achievement relative to their age and classmates. Their current performance can be compared to their prior performance, and their future growth can be charted on the same scale. The consequences of this change in the procedures employed for measurement in both education and psychology have the potential to transform both research and practice.

References

Andrich, D. 1982. Using latent trait measurement models to analyse attitudinal data: A synthesis of viewpoints. In: D. Spearritt (ed.) *The Improvement of Measurement in Education and Psychology*. ACER, Hawthorn, Victoria.

Andrich, D. 1995. Hyperbolic cosine latent trait models for unfolding defined responses and pairwise preferences. *Applied Psychological Measurement* **19**(3), 269–290.

Baker, F. B. and Al-Karni. 1991. A comparison of two procedures for computing IRT equating coefficients. *Journal of Educational Measurement* **28**(2), 147–162.

Bejar, I. I. 1983. *Achievement Testing: Recent Advances*. Sage Publication. Beverly Hills, California.

Bentler, P. M. and Houck, E. L. 1996. Structural equation modeling multiple sclerosis disease status. *The International Test Commission Newsletter,* **6**(1), 11–13.

Brick, J. M., Broene, P., James, P. and Severynse, J. 1997. *A User's Guide to WesVar*. Westat, Inc. Rockville, Maryland.

Farish, S. 1984. *Investigating Item Stability*. ACER, Hawthorn, Victoria.

Fisher, G. H. and Molenaar, I. W. 1995. *Rasch Models Foundations, Recent Developments and Applications* Springer-Verlag, New York.

Gustafsson, J.-E. and Stahl, P. A. 1996. *STREAMS User's Guide Version 1.6 for Windows*. University of Göteberg, Möendal, Sweden.

Hattie, J. 1985. Methodology review: assessing unidimensionality of tests and items. *Applied Psychological Measurement,* **9**(2), 139–164.

Hungi, N. 1997. Measuring Basic Skills Across Primary School Years. Unpublished MA Thesis. The Flinders University of South Australia.

Jöreskog, K. G. and Sörbom, D. 1993. *LISREL 8: Structural Equation Modeling with SIMPLIS Command Language*. Lawrence Erlbaum Associates, Hillsdale, New Jersey.

Lietz, P. 1996. *Changes in Reading Comprehension across Cultures and over Time*. Waxman, Munster, Germany.

Longford, N. T. 1994. *Models for Uncertainty in Educational Testing*. Springer-Verlag, New York.

Ludlow, L. H. 1994. Omitted and Not-Reached Responses. Paper Produced for the Technical Advisory Committee (TAC) of the International Association for the Evaluation of Educational Achievement, IEA, The Hague.

Luo, G., Andrich, D., and Sheridan, B. 1997. *RUMFOLD. A Computer Program*. Murdoch University, Perth.

Marsh, H. W. and Hocevar, D. 1985. Application of confirmatory factor analysis to the study of self-concept: First- and higher-order factor models and their invariance across groups. *Psychological Bulletin,* **97**(3), 562–582.

Mohandas, R. 1996. Test Equating, Problems and Solutions. Unpublished MEd thesis. The Flinders University of South Australia.

Samejima, F. 1997. Graded response model. In: W. J. van der Linden and R. K. Hambleton. 1997. *Handbook of Modern Item Response Theory*. Springer, New York

van der Linden, W. J. and Hambleton, R. K. 1997. *Handbook of Modern Item Response Theory*. Springer, New York.

Name index

The Name Index has been compiled so that the reader can proceed directly to the page where an author's work is cited, or to the reference itself in the bibliography. For each name, the page numbers for the bibliographic section are given first, followed by the page number(s) in parentheses where that reference is cited in text. Where a name is referred to only in text, and not in the bibliography, the page number appears only in parentheses.

Subject index

achievement, 1, 13, 15–16, 30, 32, 44, 62, 71, 96, 101, 111, 135, 165–166, 173, 195, 212, 218, 242, 254–255, 257, 259, 261, 280
  testing, 11, 119, 257, 260
  tests, 10, 14, 24, 30, 171, 173, 198, 200–201, 256–258, 260, 270
adaptive testing, 49, 107, 129–132, 134–136, 140, 189
American College Testing, 173
anchor item equating, 272
answer-until-correct scoring, 104
Assessment, 13, 15
attitude,
  change of, 279
  measurement of, 2, 50, 53, 103, 108, 112, 118, 277
  scales, 15, 30–31, 41, 268–269, 277
Australian Council for Educational Research, 163

bandwith, 5
bank of items, 60, 64, 89, 91, 211, 215–216

calibration, 2, 11–13, 28, 31, 34–35, 38–41, 59, 61, 71, 87–91, 93, 95, 107, 187, 193, 203, 208, 210–212, 214–217, 230–231, 269–270, 272–273, 277, 279–280
change,
  in human characteristics, 2, 279
  measurement of, 157–158
classical test theory, 8–11, 13–15, 23–25, 30–31, 44–45, 48, 55, 57, 74, 176, 195, 202, 274, 276
classroom testing, 40, 186
common item difference equating, 39
computer, 1, 3, 7, 16, 40–41, 55, 59, 62, 107–108, 136, 139, 141, 144, 147, 187, 191, 208, 212, 217, 230, 260, 268, 273–274
computerized educational testing, 115, 140
computer adaptive testing, 12, 107–108, 130, 139, 186, 205, 217, 274
conceptual understanding, 254, 257, 259, 261, 264

concurrent equating, 39–41, 273
confidence testing, 240–241
confirmatory factor analysis, 5, 31, 271–272
conjoint measurement, 10–11, 85
contingency table, 110, 112, 114–115, 117, 223–227, 229
correction for guessing, 9, 236
criterion referenced tests, 9, 196

descriptive scales, 11
development level,
  measurement of, 151, 160
diagnostic assessment, 196
differential item functioning (DIF), 12, 221, 224–232
difficulty parameter, 11–12, 37, 46–47, 77, 88, 96, 214, 239
discriminating powers, 29, 31
discrimination parameter, 12, 29, 47–48, 50, 52, 56, 74, 156, 203, 214, 239

educational measurement, 5, 7, 10, 16–17, 122, 144, 160, 176, 268
efficiency in testing, 129, 190
equating, 39
  designs, 168, 172
  of tests, 12, 53, 61, 164
errors, 11, 14, 62, 65, 88, 100, 102, 106, 135, 147–148, 180, 200, 251, 258, 268, 273–275
errors of measurement, 4, 39, 132, 200, 274–276
essay, 2, 15, 103, 152, 172, 183, 197
  scaling of, 2
estimation procedure, 3, 11, 38, 40–41, 49, 90, 132, 249, 274, 277
evaluation, 1, 3, 13–15, 74, 92, 96, 136, 138, 203, 218, 224, 232, 240

fidelity, 4–6, 31, 36–37
fit analysis, 89
fixed point, 10–11, 28, 37, 40
formative assessment, 196

gender differences, 71, 73
generalizability theory, 9–10, 244, 246